Floral Diversity and their Conservation

Floral Diversity and their Conservation

Editors
D.R. Khanna
A.K. Chopra
Rakcsh Bhutiani
Gagan Matta
Vikas Singh

Associate Editor
Chakresh Pathak

2013
BIOTECH BOOKS

ISBN: 978-81-7622-286-0

Published by : **BIOTECH BOOKS**
4762-63/23, Ansari Road, Darya Ganj,
New Delhi - 110 002
Phone: +91-011-23262132
E-mail: biotechbooks@yahoo.co.in

Laser Typesetting : **Sushil Kumar**

Printed at : **Chawla Offset Printers**
Delhi - 110 052

PRINTED IN INDIA

Preface

Over the past two decades, there has been increasing interest in understanding how species diversity affects the functioning of ecosystems. Research in this area has often been justified on grounds that loss of biological diversity ranks among the most pronounced changes to the global environment and reductions in diversity and corresponding changes in species composition could alter fluxes of energy and matter that underlie important services that ecosystems provide to humanity (e.g., production of food, pest/disease control, water purification).Plants are universally recognized as a vital part of the world's biological diversity and an essential resource for the planet. Many thousands of wild plants have great economic and cultural importance, providing food, medicine, fuel, clothing and shelter for humans around the world. Plants also play a key role in maintaining the Earth's environmental balance and ecosystem stability. They also provide habitats for the world's animal and insect life.Many plant species are threatened by habitat transformation, over-exploitation, invasive alien species, pollution and climate change, and are now in danger of extinction. The disappearance of such vital and large amounts of biodiversity presents one of the greatest challenges for the world community: to halt the destruction of plant diversity that is essential to meet the present and future needs of humankind.

The present book provides an introduction to the incredible diversity of the plants. The book is designed to incorporate both new and traditional methods to provide a foundation in the study of the diversity of land plants from a phylogenetic perspective.

The present work is the compilation of various papers contributed by learned community from different part of the World working in the field of plant diversity conservation this book describes the unpublished research of the author.

However, we are confident that this book will serve as a valuable communication and awareness-raising tool for the further implementation of the plant conservation strategy.It is also hoped that in the present scenario, when in every part of the world the emphasis is giving on Plant diversity conservation this volume will be of immense use, particularly in opening new vistas of the wonderful world of Plant diversity.

At last but not least we thank all our contributors who have devoted their precious time in contributing their research in the form of papers for this volume. The thanks are also due to M/s Biotech books publishing house, New Delhi in bringing out this volume in time with nice presentation.

Editors

Contents

Floral Diversity and their Conservation (2013), Editors: D.R. Khanna et al.
Published by Biotech Books.
ISBN: 978-81-7622-286-0 Pages: 1-11

1

Some Ethno-medicinal plants used for family planning and treatment of sexual disease in southern Rajasthan, India

B. L. Yadav, Gajendra Singh and Prakash Chandra Aheer
Department of Botany
M. L. V. Government College, Bhilwara

An extensive survey was conducted by frequent planned visits to different parts of southern Rajasthan during 2005 – 2010 to document the information on medicinal plants which are being used by the ethnic groups in this part of the state. The study is based on the exhaustive interviews with local physician practicing indigenous system of medicine, village headman, bhopa, shepherds, birth attendants and tribal folks. During the study 24 plant species belonging to 21 families used in herbal remedies by tribes of southern Rajasthan to treat ailments such as syphilis emmenogogue, fertility, vigor & vitality, galactogogue, irregular menses, menstrual disorders, sexual diseases, swelling of testis, to check bleeding after delivery and to control birth rate using fresh as well as dried plant material have been recorded. An enumeration of plant species with the botanical name, local name, plant parts used and mode of administration for effective control of different sexual disorders is presented in this paper.

Key words: Ethno-medicine, sexual disorders, tribals.

INTRODUCTION

India is one of the 12 mega-diversity countries of the world and is a repository of medicinal plants. Rajasthan is one of the largest states and is located in the North western part of India. Geographically, it lies between 23°3' to 30°12' N longitudes and 69°30' to 78°17' E latitudes. Southern part of Rajasthan comprising Banswara, Chittorgarh, Dungarpur, Pratapgarh, Rajsamand, Sirohi and Udaipur districts (Fig.1) is the tribal belt in which *Bheel, Damor, Garasia, Kathodia* and *Meena* are the main tribes. For these people, the surrounding plants form an integral part of their culture and informations regarding their uses are passed on orally from generation to generation.

The tribals of remote areas of southern Rajasthan are totally dependent on herbs for their healthcare (Jain *et al.,* 2005) because of lack of sufficient medical facilities and roads in the remote areas and the cost of medicines. Studies on ethno-medicine have been carried out in Rajasthan by various botanists from different parts of the state (Jain *et al.,* 2005, 2008; Joshi, 1995; Katewa & Arora, 1997; Katewa & Galav, 2006; Katewa *et al.,* 2003, 2008; Meena & Yadav, 2007, 2010; Sharma, 2002; Sebastian & Bhandari, 1984, 1988; Singh, 1999; Singh & Pandey, 1980; Trivedi, 2002)

The southern part of Rajasthan is characterized by the tropical deciduous type of vegetation consisting of *Anogeissus latifolia* (Roxb. ex DC.) Wall. ex Guill & Perr., *Anogeissus pendula* Edgew, *Balanites aegyptiaca* (Linn.) Delile., *Boswellia serrata* Roxb., *Diospyros melanoxylon* Roxb., *Madhuca indica* J.F. Gmelin, *Tectona grandis* Linn. f., *Terminalia arjuna* (Roxb. ex DC.) Wight & Arn. etc. as the important plant species.

Ethno-medicine is the system maintaining health and curing diseases based on folk beliefs and traditional knowledge, skills, methods and practices. The local tribals harbor the vast diversified flora. Some of the herbal medicines are being used for population control and to treat sexual diseases. Traditional method based on ethno-medicine is used to control population growth rate which includes abortion at initial weeks or preventing conception. The tribal people also use the local herbal medicine to cure sterility, enhance the chances of conception and to cure sexual diseases like leucorrhoea, gonorrhoea and to regularize menses.

The aim of the present study was to document the ethno-medicinal plants used to control sexual diseases and birth rate.

MATERIALS AND METHODS

Field trips in different parts of southern Rajasthan were conducted with the local medicine men. Before launching the field trips, it is essential to establish the rapport with the tribals as generally tribals, who know about the herbal medicine, do not want to disclose the plant because they believe that when identity of the medicinal plant is disclosed its medicinal properties will be lost for ever. Secondly they believe that the knowledge is exclusively for their community and they feel that they are being cheated by the outsiders. The study is based on the exhaustive interviews with local physicians practicing indigenous system of medicine, village headman, bhopa, shepherds, birth attendants and tribal folks. For authenticity about medicinal properties of plants the information collected during fieldwork were verified at different places through different informants and in different seasons.

Each of the plant species recorded has been collected with the help of the informants and photographs were taken. The species were identified with the help of reputed flora of India Series- 2 (Shetty & Singh, Vol. 1987-1993). The voucher specimens were deposited in the Herbarium, Department of Botany, MLV Government College, Bhilwara.

ENUMERATION

The data on ethnomedicinal plants such as the botanical name, local name, family, plant parts used and the traditional methods of drug administration in different ailments are being presented in **table 1**. In the enumeration, data on medicinal plants are arranged alphabetically, each by its botanical names followed by name of family and local names.

During the present study, 24 plant species belonging to 21 families being used by tribes of southern Rajasthan to treat ailments such as to control emmenogogue, fertility, vigour & vitality, galactogogue, irregular menses, menstrual disorders, sexual diseases, swelling of testis, to check bleeding

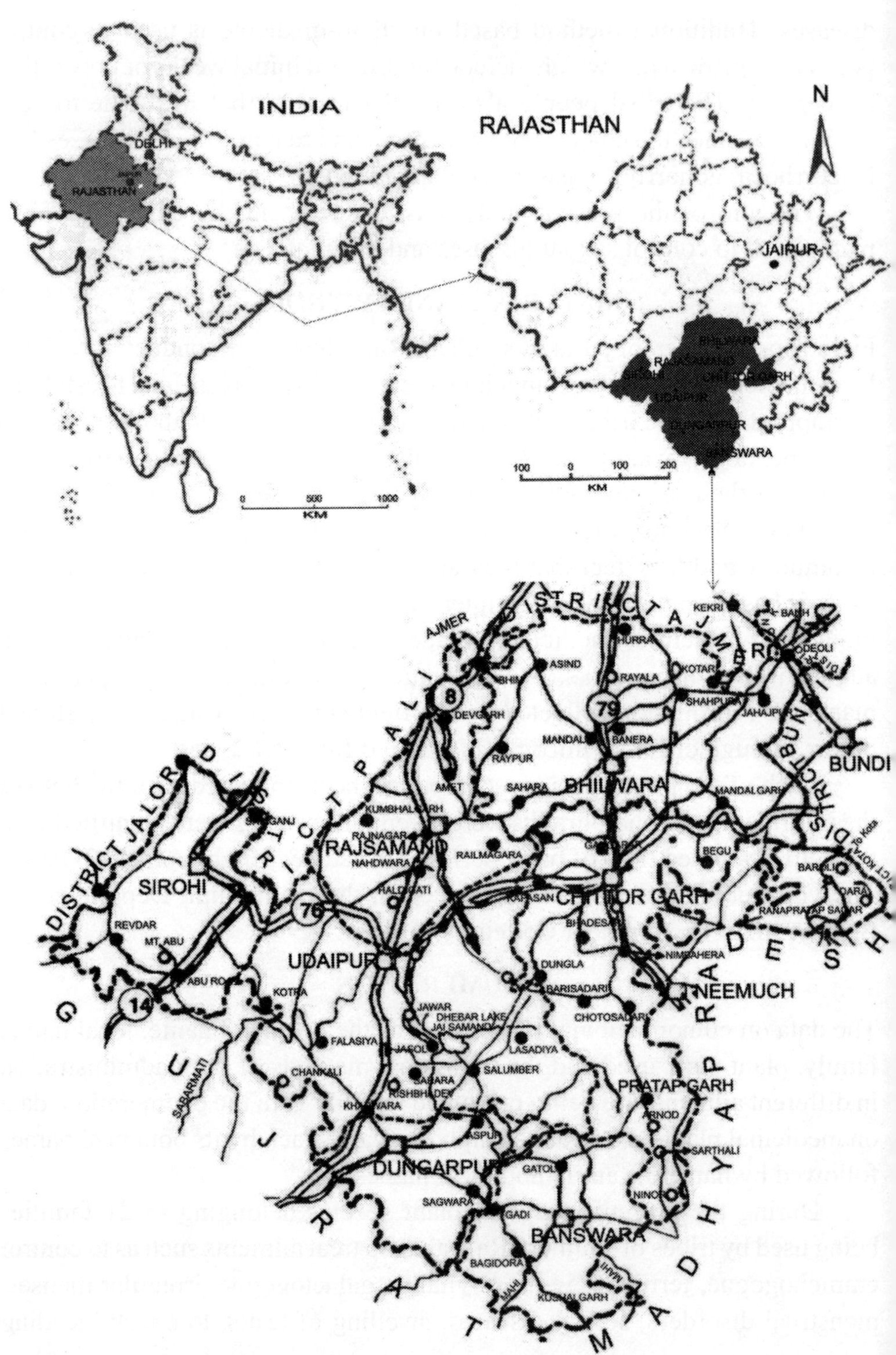
INDIA
DELHI
RAJASTHAN
0 500 1000
KM
RAJASTHAN
N
JAIPUR
BHILWARA
RAJASAMAND
CHITTOR GARH
UDAIPUR
DUNGARPUR
BANSWARA
100 0 100 200
KM
DISTRICT AJMER
AJMER
KEKRI
HURRA
ASIND
RAYALA
KOTARI
DEOLI
8
79
DEVGARH
MANDAL
BANERA
SHAHPURA
JAHAJPUR
RAYPUR
DISTRICT BUNDI
BUNDI
SAHARA
BHILWARA
MANDALGARH
AMET
DISTRICT PALI
KUMBHALGARH
RAJNAGAR
RAJSAMAND
NAHDWARA
RAILMAGARA
BEGU
DISTRICT JALOR
SIROHI
REVDAR
MT. ABU
76
UDAIPUR
KAPASAN
CHITTOR GARH
BHADESAR
RANAPRATAP SAGAR
NIMBAHERA
DUNGLA
BARISADARI
NEEMUCH
14
KOTRA
JAWAR
DHEBAR LAKE
JAI SAMAND
LASADIYA
CHOTOSADARI
FALASIYA
CHANKALI
SALUMBER
SABARMATI
PRATAP GARH
ARNOD
SARTHALI
DUNGARPUR
SAGWARA
BANSWARA
BAGIDORA
KUSHALGARH
MAHI
GUJRAT
MADHYA PRADESH

Table 1: Some ethno-medicinal plants used for family planning and treatment of sexual disease in southern Rajasthan.

S.No.		*Family*	*Local name*	*Mode of administration*
4.	*Amaranthus viridis* L.	Amaranthaceae	*Dandi*	10 gm root crushed to prepare 50 ml of decoction and used thrice a day to cure menstruation problems. The 5 gm seed paste is applied on back bone during pregnancy to cure backbone ache.
5.	*Asparagus racemosus* Willd	Liliaceae	Satawari	1 tsp juice of tuberous roots is given orally for 15 days to cure sexual debility.
6.	*Boerhavia diffusa* L.	Nyctaginaceae	*Punernava*	1 tsp. leaves juice is taken 4 - 5 times a day orally to check bleeding after delivery.
7.	*Bombax ceiba* L.	Bombacaceae	Sibal	Half cup decoction of stem bark of *Bombax ceiba* L. mixed with ½ cup decoction of *Abelmoschus esculentus* (L.) Moench. root is taken orally for 8 days by males, it regenerates the fertility. 5 gm of flower powder is taken by women with milk to cure menstrual disorders.
8.	*Boswellia serrata* Roxb. ex. Cocls	Burseraceae	Salar	1 kg gum is taken with sweet dish for 15 days once a day to cure impotency in men.
9.	*Butea monosperma* (Lam.) Taubert.	Papilionaceae	*Khankra*	Gum is known as *kamarkas*. It is used in preparation of *Laddu* (A mixture of wheat flour, gum and sugar) and is given after delivery.
10.	*Carica papaya* L.	Caricaceae	End kakari	200 gm of unripe fruits are used orally once a day for three days in early stages of pregnancy for abortion.
11.	*Ceropegia bulbosa* Roxb.	Asclepiadaceae	*Khadula*	Two raw tubers are eaten daily during rainy season by male and female to enhance the fertility & vitality.

contd...

S.No.		Family	Local name	Mode of administration
12.	*Chlorophytum tuberosum* (Roxb.) Baker	Liliaceae	*Dholi musali*	500 gm dried fasciculated roots are used in the preparation of *laddu* with 1 kg dried gum of *Anogeissus latifolia* wich are taken for 15 days during winter season by male and female to enhance the fertility, vigour & vitality.
13.	*Costus speciosus* (Koen.) Sm.	Costaceae	Mahalakari	1 tsp. root powder is taken by male twice a day for 3 - 4 days in sexual debility.
14.	*Echinops echinatus* Roxb.	Asteraceae	Oont Katilo	Pieces of 5 cm size of its fresh roots are kept at the back of head touching scalp or coil the fresh rooys and kept on naval before parturition time or during labour pains for easy delivery
15.	*Ehretia laevis* Roxb.	Ehretiaceae	Tambolan	Five to ten ml bark juice is given orally just after delivery, it relives delivery pain.
16.	*Erythrina stricta* Roxb.	Papilionaceae	Kesuri	Twenty to thirty ml juice of fresh or dried flowers is given orally for abortion; abortion takes place within 12 hrs only one dose is sufficient.
17.	*Ficus bengalensis* L.	Moraceae	*Bar, Bargad*	Five or ten drops are taken with sweet (*patasa)* by male for 45 days to make seamen thick and regain sexual potentiality.
18.	*Gloriosa superba* L.	Liliaceae	Kalahari	1 tsp extract of tuber is mixed with root extract of *Solanum virginianum* L. and taken by female as abortifacient.
19.	*Habenaria marginata* Colebr. (Fig. 6)	Orchidaceae	Piwali	10 gm tuber paste is applied externally to cure, swelling of testis.
20	*Leea macrophyla* Roxb. ex Hoenem.	Leeaceae	Hastipalash	½ - 1 tsp root powder is taken by male once a day for 5 - 7 days to cure sexual debility.

contd...

S.No.		Family	Local name	Mode of administration
21.	*Tecomella undulata* (Roxb.) Seem	Bignoniaceae	Rohira	1 gm of bark powder is taken with 100 ml of hot milk for 7 days for abortion.
22.	*Terminalia arjuna* (Roxb. ex DC.) Wight & Arin	Combretaceae	Kahua	1 gm tender leaf paste mixed with 2 tsp sugar and 100 ml of cow milk. It is given once a day for 21 days to cure Pspermatorrhoea.
23.	*Trachyspermum ammi* (L.) Sprague ex Terrill	Apiaceae	*Ajwain*	1 kg of seed powder mixed with *Gur* and *Desi ghee* (Butter) is taken orally once a day to treat and regularize scanty menuatruation. It is also used to clear uterus and regularise menstrual cycle
24.	*Withania coagulance* (Stocks.) Dunal	Solanaceae	*Paneer bandh*	10 gm fresh leaf paste applied on testis to get relief from swelling and pain. Fruit infusion is used by females as emmenogogue and galactogogue.

after delivery and to control birth rate using fresh as well as dried plant material have been recorded. They firmly believe in the traditional way of treatment of various ailments using medicinal plants rather than modern medical treatment.

DISCUSSION

Much of the world's population depends upon traditional medicine to meet daily health requirements (Eddouks *et al.,* 2002). During the present study 24 plant species belonging to 21 families which are being used as ethno-medicine to control sex related diseases and to check birth rate by the tribals of southern Rajasthan have been recorded. Of these, 3 families are of monocots, 16 are of dicots and two are of pteridophytes. *Carica papaya* L., *Gloriosa superba* L., *Erythrina stricta* Roxb. and *Tecomella undulata* (Roxb.) Seem are usually used as abortifacient to control birth rate in initial stages, while *Amaranthus viridis* L. is used to regularize menstrual cycle. The tribal people use *Habenaria marginata* Colebr. and *Withania coagulance* (Stocks.) Dunal to get relief in swelling on testis. *Chlorophytum tuberosum* (Roxb.) Baker is used to enhance sexual potency in males. *Butea monosperma* (Lam.) Taubert. and *Trachyspermum ammi* (L.) Sprague ex Terrill are used for regularizing menses after delivery. These findings have not been reported by earlier worker from Rajasthan.

Some plants such as *Chlorophytum tuberosum* (Roxb.) Baker, *Ceropegia bulbosa* Roxb. and *Ficus benghalensis* L. are used in recovery of fertility in both male as well as female. The roots of *Echinops echinatus* are kept at the back of head for easy delivery which was earlier reported by Meena & Yadav (2010). *Abutilon indicum* (L.) Sweet. is used to treat syphilis which has also been reported by Muhammad & Khan (2008) from Pakistan.

The traditional knowledge is fast being depleted due to modernization and diminishing interest of younger generation. There is an urgent need to document this precious knowledge for the future as this knowledge is propagating through oral means only (Jain, 1967). The traditional uses of herbal remedies has further declined due to scarcity caused by human activity, drought and overgrazing by domestic animals threatening the plant diversity. Therefore the conservation of medicinal plants require immediate steps to be taken up (Katewa & Galav, 2006)

The exploitation of some of the species like *Costus speciosus* (Koen.) Sm., *Habenaria marginata* Colebr., *Leea macrophyla* Roxb. ex Hoenem. and *Withania coagulance* (Stocks.) Dunal for ethno-medicine in southern Rajasthan is unsustainable and might threaten the local plant population. Afforestation, protection and cultivation of precious wild medicinal flora of an area are necessary steps for long time use.

CONCLUSION

There is much ethnomedicinal knowledge concerning sexual disease and birth control treatment within tribal communities of southern Rajasthan. Its documentation should be made on priority basis before it is lost for ever. Ethno-medicinal knowledge of tribals of the area on abortion, sexual fertility and female contraceptives, which is one of the important information innovations used by tribals, is quite relevant to present day situations. The other issues needed to be addressed are efficacy, quality, safety and standardization of doses. The ethno-medicinal information require further research, while efficacy of various indigenous practices and folklore uses should be subjected to pharmaceutical and phyto-chemical investigations in order to identify how these can be of practical advantage in medicine development.

ACKNOWLEDGEMENTS

Authors are highly thankful to all the tribal informants - Shri Gobaru Ji, Jetha Ram Ji, Laxman Ram Ji, Limba Ram Ji, Modan Ji, Nana Ram Ji, Pratap Singh Ji, Ram Chandra Ji, Salira Ram Ji and Vikram Ji for their cooperation and help during present ethnobotanical study by way of providing traditional knowledge. Help rendered by district forest officer Chittorgarh, Sirohi, Pratap garh Banswara and Dungerpur for providing the facilities during the field work. Thanks are also due to Shri Bhanwar Singh Ji Meena, Additional Superintendent of Police, district Sirohi for his cooperation during these studies. Authors are thankful to Principal and Vice Principals of MLV Government college, Bhilwara for providing the facilities. Financial assistance provided by UGC Bhopal is gratefully acknowledged.

REFERENCES

1. Eddouks M, Maghrani M, Lemhadri A, Ouahidi M L & Jouad H 2002 Ethnopharmacological survey of medicinal plants used for the treatment of diabetes mellitus, hypertension and cardiac disease in the southern region of Morocco (Tafilalet). *J Ethnopharamacol* 82: 93.
2. Jain S K 1967 Ethnobotany: Its scope and study, *Indian Museum Bull*, 2: 39-43.
3. Jain A, Katewa S S, Galav P 2005 Some phytotherapeutic claims by tribals of southern Rajasthan. *Indian J Traditional Knowledge*, 4(3): 291 - 297.
4. Jain A, Katewa S S, Galav P, Nag A 2008 Some therapeutic uses of biodiversity among the tribals of Rajasthan, *Indian J Traditional Knowledge*, 7 (2): 256-262.
5. Joshi P 1995 Ethnobotany of the Primitive Tribes in Rajasthan, Printwell, Jaipur
6. Katewa S S & Arora A 1997 Some plants of folk medicine of Udaipur district, Rajasthan, *Ethnobotany*, 9: 48 - 51.
7. Katewa S S, Chaudhary B L, Jain A & Galav P K 2003 Traditional uses of plant biodiversity from Aravalli hills of Rajasthan. *Indian J Traditional Knowledge*, 2 (1): 27 - 39.
8. Katewa S S, Galav P K, Nag A & Jain A 2008 Poisonous plant of the southern Aravalli hills of Rajasthan, *Indian J Traditional Knowledge*, 7 (2): 269- 272.
9. Katewa S S & Galav P K 2006 Additions to the traditional folk herbal medicine from Sekhawati region of Rajasthan. *Indian J. Traditional Knowledge* 5(4): 494 - 500
10. Meena K L & Yadav B L 2007 Some ethno-medicinal plants of Rajasthan. In P. C. Trivedi (Ed.) *Ethnomedicinal Plants of India*. Aavishkar publishes Distributors, Jaipur, 33 - 44.
11. Meena K L & Yadav B L 2010 Some ethno-medicinal plants of southern Rajasthan. *Indian J Traditional Knowledge*, 9 (1): 169 - 172.
12. Muhammad I C & Khan M A 2008 An ethnomedicinal inventory of plants used for family planning and sex diseases in Samahni valley, Pakistan 7(2): 277-288.
13. Sebastian M K & Bhandari M M 1984 Medico-ethno-botany of Mt. Abu, Rajasthan, *J Ethnopharmacol*, 12: 223 - 230 .
14. Sebastian M K & Bhandari M M 1988 Medicinal plant lore of Udaipur district Rajasthan, *Bull Med Ethnobot Res*, 5 (3-4): 133 - 134.

15. Sharma N K 2002 Ethno-medico-Religious plants of Hadoti Plateau (S E Rajasthan) - A Preliminary Survey In P. C. Trivedi (Ed.) *Ethnobotany,* Aavishkar publishes Distributors, Jaipur, 394 - 411.

16. Singh G 1999 A contribution of ethno-medicine of Alwar district of Rajasthan, *Ethnobotany*, 11: 97 - 99.

17. Singh V & Panday R P 1980 Medicinal plant lore of the tribals of eastern Rajasthan, *J Econ Taxon Bot*, 1: 137 - 147.

18. Shetty B V and Singh V 1987 - 1993 *Flora of Rajasthan,* Vol. I - III BSI, Howrah

19. Trivedi P C 2002 Ethno-medicinal plants of Rajasthan State India, In P. C. Trivedi (Ed.) *Ethnobotany,* Aavishkar publishes, Distributors, Jaipur 412 .

Floral Diversity and their Conservation (2013), Editors: D.R. Khanna et al.
Published by Biotech Books.
ISBN: 978-81-7622-286-0 Pages: 13-17

2

Addition to the Myxomycetes Flora of India

Jadhav D. M. and S. P. Nanir*
Deptt. of Botany, G.M.D. Arts B.W. Commerce & Science College, Sinnar
*Dept of Botany, Institute of Science, Caves Road, Aurangabad

The Genus ***Diderma*** Pers. is represented by over 57 species from the world. Till now about 29 species have been described from the Indian flora and about 11 species from the state of Maharashtra. In the present paper 2 species are being described and illustrated from the region of north eastern ranges of western Ghat and constitute the addition to the list of Myxomycetes flora of India. *viz.* ***Diderma donkii*** Nann. –Brem. and ***Diderma lohogadensis*** Patil Ranade and Mishra. Former species constitute new record to the Indian Myxomycetes while later has been collected for first time after type collection.

Key words: Fungi, Myxomycetes, Diderma

DIDERMA DONKII NANN.-BREM

Fructification predominantly sessile, sporangiate often associated with short plasmodiocarp, scattered, pale yellow or chalky white. Sporangia depressed mm .in diam. and 0.09-1.2 mm. in height, concave above pulvinate with dark brown shiny wrinkled base. Plasmodiocarp simple, dichotomously branched, depressed and concave above, dark brown and wrinkled towards the base, pale yellow or chalky white. Hypothallus distinct,white, with granular lime. Peridium double, outer layer thick, brittle, impregnated with lime granules, pale yellow, dark shiny brown towards the base, inner layer thin, membranous, ash gray, hyaline under transmitted light; dehiscence circumssile or by upper slit, lower part persistent as a saucer shaped structure. Columella represented by raised base, rough, pale yellow, limy with granular lime. Capillitium abundant elastic, filamentous, smooth, dichotomously branched, anastomosed with cross bars and perforated membranous expansion at the dichotomy, small perforation towards the apices, hyaline throughout, tips are attached to peridium. Spore mass brown, violaceous brown under transmitted light, globose, 6.0 – 7.5 (-9.0) µm in diam., warty, warts are in cluster as well as in straight and curve line. COLLECTION EXAMINED: SPN.DMJ./ 2402, Sept.1992, Saputara, on dry angiospermic plant.

The present species is characterized by depressed pulvinate sporangia associated with small simple plasmodiocarps with dark brown peridium towards the base of fruiting body capillium with perforated membranous expansion at the dichotomy and small perforation towards the apices and conspicuously warted spores. The species is similar to type description. However it differs in its depressed sporangia associated with small flat pulvinate plasmodiocarpous habit, dark brown peridium towards base of the fruiting, capillitium with membranous expansion and conspicuous warted spors.

The species have similarities with ***D. spumaroids*** (Freis), but differs from it in its mostly distinct peridial layer and in the shape and colour of sporangia. In shape it looks more like ***D. testaceum*** *(*Schrad) Pers. It differs from ***D. globosum*** in its shape and colour but similar in spore character. The species constitute the new record to the list of Indian myxomycetes.

DIDERMA LOHOGADENSIS PATIL, RANADE & MISHRA

Fructification sporangiate, stipitate, gregarious, reddish to chestnut brown, 1.0-2.6 mm. tall Sporangia infundibuliform, cup dark brown, 0.85-1.43 mm. in diam. Stipe thick, stout, erect or bent, cylindrical, rugose, yellowish brown to whitish towards the bases reddish brown to chocolate brown towards the apex, filled with lime nodules, 0.71-1.7 mm. long and 0.46-0.54 mm. thick. Hypothallus rotate, shining, white to ochraceous, with lime nodules depositions. Peridium triple; outer layer thick, cartilaginous, limy, impregnated with lime granules smooth, orange, adhered to middle layer; middle layer thick, whitish or orange limy; inner layer remote, grayish white sprinkled with lime granules on outer surface where as inner surface is brown with irregular patches, thin, membranous; dehiscence irregular, some time outer two layer separate as a cup, inner layer persists. Capillitium abundant, thin, filamentous, slender, branched, sparsely anastomosed, attached at both the ends, filaments smooth, brown, with expanded and perforated hyaline ends. Collumella absent. Spore dark brown to black in mass, deep reddish brown to pinkish under transmitted light, globose to sub globose, 10-13.5 (-15) µm. in diam. Paler on one hemisphere, minutely and uniformly warted or spinulose, warts unequal in length, shorter at paler side, arranged in small lines.

COLLECTION EXAMINED: SPN.DMJ/2046, Aug.1990, Vani, Nashik 2195, 2196, 2336, 2337, Sept.1991, Neelamati, Thane; 2338, Aug.1991, Bhandardara, Ahemadnagar.On dry leaves and stem of angiospermic plant.

The species can be characterized by its quite unique shape of sporangia. The species has been collected for the first time after type collection.

ACKNOWLEDGEMENTS

Our thanks are due to the Principal Dr. R. N. Bhaware G. M. D. Arts B.W. Commerce & Science College, Sinnar for providing facilities.

Diderma donkii Nann.-Brem. Habit 35 x

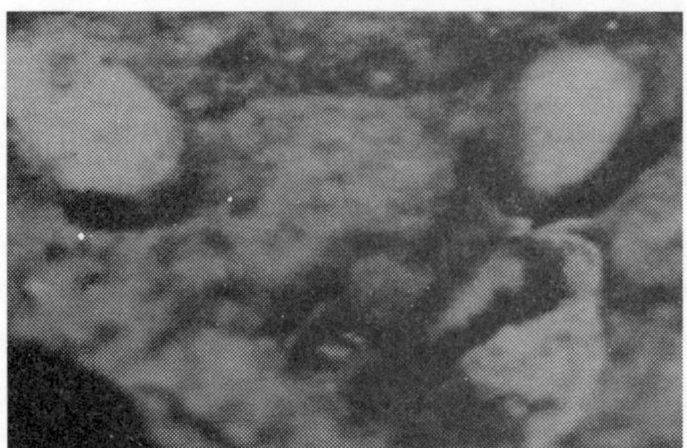

Dehisced Sporangia 30x

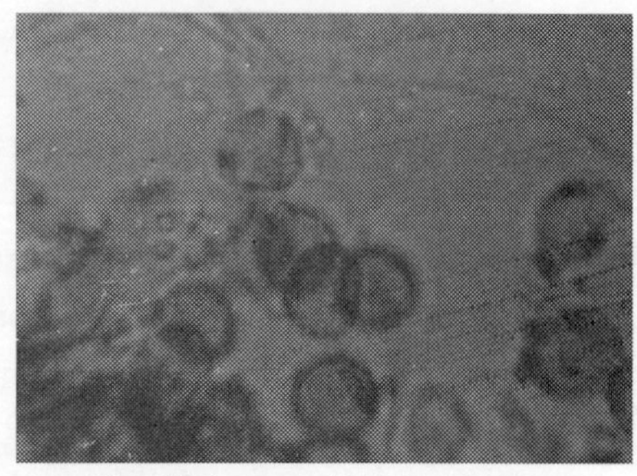

Capillitium and Spores 1000 x

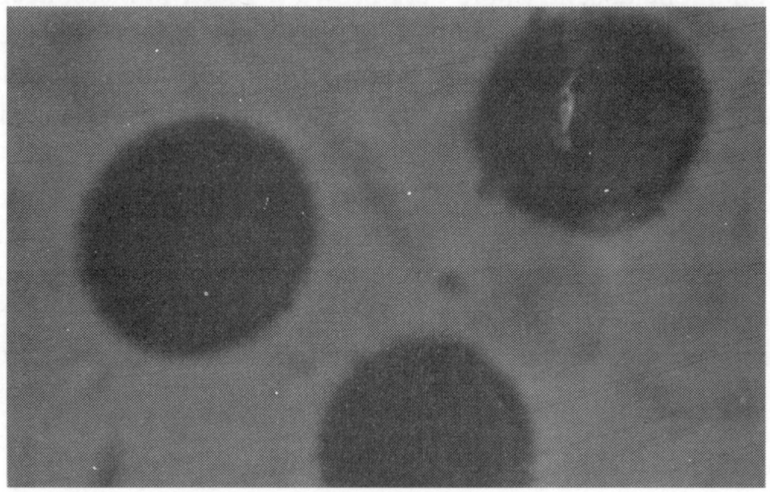

Spores 1000 x

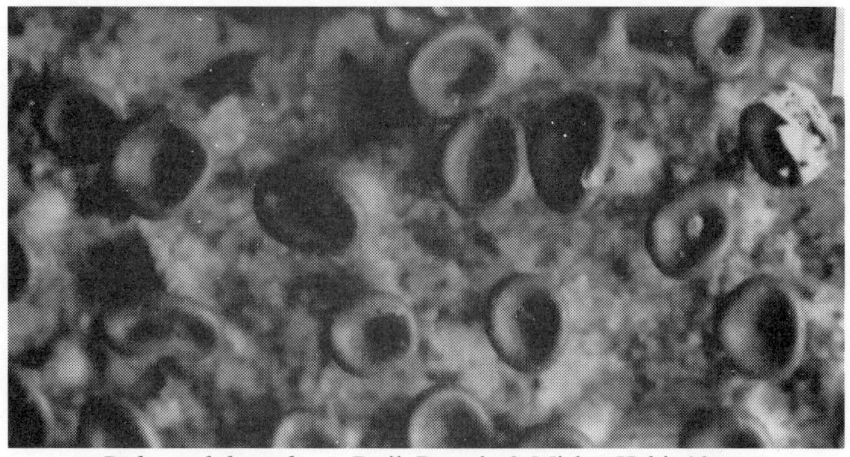

Diderma lohogadensis Patil, Ranade & Mishra Habit 10x

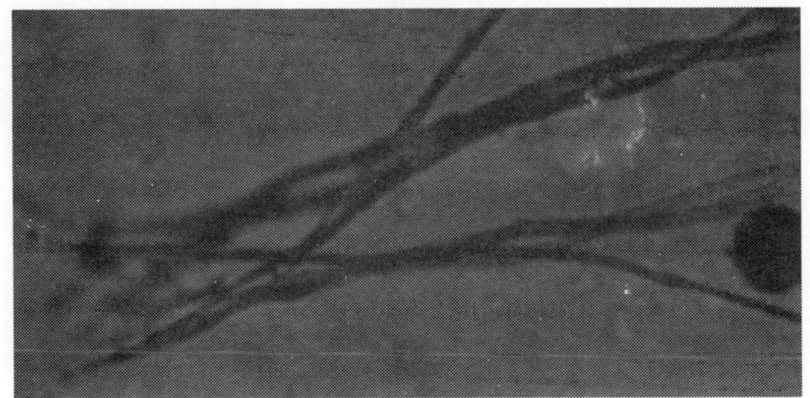

Capillitium 400x

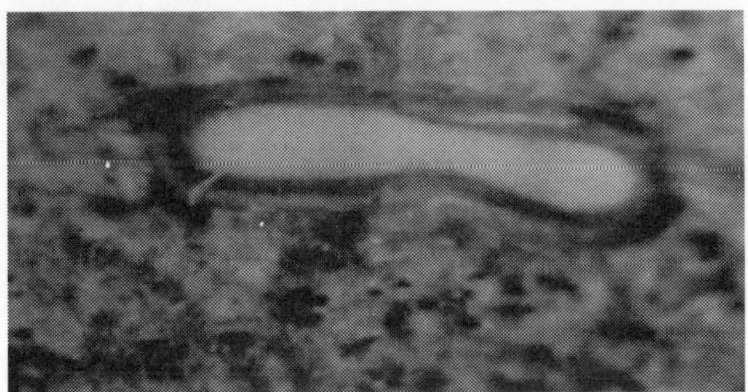

Collumella 40x

REFERENCES

1. Farr, M .L. (1976): Flora Neotropica, Mon, 16 Myxomycetes . The New York Botanical Garden. N. Y.
2. Lakhanpal, T.N. & K. J. Mukarji (1981) Indian Myxomycetes, J.C.Carmer.
3. Lodhi S.A. (1934): Indian Slime Moulds(Myxomycetes) Punjab University, Lahore.
4. Martin. C. W. and C. J. Alexopoulos (1969) : The Myxomycetes, Univ. of Iowa Press Iowa City.
5. Martin. C.W. and C.J. Alexopoulos and M. L. Farr (1983) Genera of Myxomycetes, Univ. of Iowa Press Iowa City.
6. Nannenga- Bremekamp, N.E. and Y. Yamamota. (1987): Addition to the Myxomycetes of Japan- II Proc. K.Ned. A. kad .Wet Ser.C, 90, 311-349.
7. Nannenga- Bremekamp, K. Ned. Akad .Wet C. 91, 416.
8. Thind K.S (1977): The Myxomycetes of India. I.C.A.R. New Delhi

Floral Diversity and their Conservation (2013), Editors: D.R. Khanna et al.
Published by Biotech Books.
ISBN: 978-81-7622-286-0 Pages: 19-33

3

Seed Development in *Euphorbia splendens* Boj

L.P. Dalal
Associate Professor, Department of Botany,
J. B. College of Science, Wardha (M.S)

An experiment was conducted to study the embryological characters in *Euphorbia splendens*. The plant shows a variety of characters like the centrally placed naked female flower of the cyathium is represented by a superior tricarpellary, trilocular gynoecium mounted on a long pedicel and surrounded by five groups of male flowers. Each locule has an anatrapous ovule. The obturator is an interesting feature in Euphorbiaceae, placental in origin and consist of loosely elongated cells. The distal part of the obturator is reported on the exostome. The endosperm development is ab-initio nuclear type. In mature seed, the embryo sac is completely filled with cellular endosperm. A thick saucer shaped pad of tissues below the chalazal end of the embryo sac was reported i.e the hypostase was a flat saucer shaped structure. No reserve food material was observed in the cells of the hypostase. The embryo development follows the Euphorbia variation of Onagrad type and falls under Grand period-I, Series A-A_2 and Megachetype IV. The seed coat develops from both the integuments and the outer epidermis of the inner integument forms the characteristic brittle stony layer of the seed coat. Hence, the family is heterogenous in nature as suggested by Hutchinson (1973).

Key words : *Euphorbia splendens* Boj. Development, Seed, Endosperm, Outer integument, Inner integument, Seed coat, Mature embryo.

INTRODUCTION

Plants and flowers are the beautiful ornaments of the nature,which gives silence to the soul and freshness to the mind. Floral parts attract insects, nematodes, animals and humans too, particularly embryologist. Embryology is the study of microsporogenesis and megasporogenesis, gametogenesis, gametophyte development, fertilization and development of endosperm, embryo development and seed coats. The possibility of utilizing embryological characters in taxonomy was indicated by some earlier botanist like Hofmeister and Strasberger. The role of embryology in solving the problems was first brought into prominence by a German embryologist, Schnarf in 1931. Embryological characters, being less prone to adaptive stress, are relatively stable and have acquired great significance in plant taxonomy, especially when external morphological characters are inconclusive and misleading as a result of convergence (Kapil and Bhatnagar, (1980). Thus embryological evidences have been used in solving the taxonomical problems at almost all levels, and have helped in resolving doubtful systematic positions of several taxa. However, according to Jones and Luchsinger,(1987), the embryological evidences have proved to be of significant help in determining relationships within families, genera and species, but less useful at the rank of order, subclass or class. Maheshwari, (1950) and John, (1963), on the basis of their extensive work on embryology, have provided list of families, tribes, etc. where embryology has either supported earlier classification or has proposed a new systematic position for the taxon concerned. In later years, similar excellent work on the role of embryology has been done by Johanson (1950), Rao, (1962), etc. These plant taxonomists have utilized embryological character's in various ways. The family Euphorbiaceae, one of the heterogenous family of angiosperms, has attracted the attention of the morphologists, cytologists, embryologists and others due to variations in characters. The family is represented by herbs, shrubs and trees. *Havea brasiliensis,* is well known tree of Amazonian forests. Many species of *Tragia* are climbers. The leaves are alternate, opposite or whorled. In a few succulents, they are reduced to spines. Euphorbiaceae is also highly diverse in anatomy and pollen morphology, yet the family resists all attempts to fragment it. Before reviewing taxonomic history of Euphorbiaceae, we have to see that the family has been given a very different status by different systematists.

Benthem and Hooker, (1883) placed this family in the cohort uni-sexuales along with 8 other families *viz.* Balanopseae, urticaceae, Plantanceae, Leitneriaceae, Jugalandaceae, Myricaceae, Casuarinae, and Cupulifae. The system of Bentham and Hooker was, however, not intended to express the phylogenetic relation of these families.

Rendle,(1925,1963), includes this family under eight order Tricoceae in Grade B-Dialypetalae. Later he followed the classification of Pax and Hoffman,(1931) in Engler and Prantle " Die Naturlichen Pflanzenfamilien", where they include the Euphorbiaceae in suborder Triccoceae of the order Geraniales.

Hutchinson,(1959) raised the family Euphorbiaceae to the status of an order Euphorbiales. He does not divide the family into any tribes. He placed this order, Euphorbiales after Malphigiales and is of the opinion that Euphorbiales which consist of a single family i.e "Compositae (heterogenous) family", is probably derived from several stocks like Bixales, Tiliales, Malvales, Celastrales and probably Sapindales.

Cronquist,(1968) in his system of classification has raised Euphorbiaceae to the order Euphorbiales like Hutchinson,(1959) under the subclass Rosideae of Magnoliatae.

Later in 1982, Cronquist modified his classification as compared to that in 1968. The Euphorbiaceae is placed in the subclass resideae, which has an order Euporbiales with 4-families, viz; Euphorbiaceae, Pandanaceae, Buxaceae, and Simmondsiaceae. The order is dominated by the very large family Euphorbiaceae, to which, the other 3-families are attached as small satellites. Of these, only the Pandanaceae is without question closely allied to Euphorbiaceae.

From the above taxonomical description, it is evident that there is no unanimous opinion among the taxonomists with regard to the status of Euphorbiaceae. Some regarded it as a family while others treat it as an order. Moreover the opinion is divided on its further division and composite nature.

Rendle,(1925,1963) include this family under eight order Tricoceae in Grade B Dialypetalae. Later he followed the classification of Pax and Hoffman,(1931) in Engler and Prantle "Die Naturlichen Pflanzenfamilien" where they include the Euphorbiaceae in suborder Tricoceae of the order Geraniales.

MATERIALS AND METHOD

The material was collected during morning and evening hours on bright sunny days and was fixed in formalin-aceto alcohol followed by preservation in 70% alcohol after 24hrs. The young cyathia were pre-treated with para-dichlorobenzene for half an hour before fixation and these pre-treated material were also preserved in 70% alcohol after 24 hours of fixation. They gave better results during study of embryology. Seeds and hard fruits were pre-treated with 10-15% KOH solution for 24 hours after fixation and pre-treated seeds were also preserved in 70% alcohol.

For embryological studies, processing, dehydration and embedding in paraffin were followed by routine methods and section were cut at 10-12 micron thick on rotary as well as rocking microtomes. Sections were stained with 0.5% iron alum haematoxylin and destained in saturated solution of picric acid. Other stains like Delafields haematoxylin and counter staining techniques by using saffranin and light green were also tried and gave better and excellent results.

KOH treated seeds were used for dissecting mature embryo and maceration of seed coat to study the sclereids. For maceration, Schultz maceration fluid was used (Johansen,1940).

Diagrams were drawn by using prism type of camera-lucida on a calibrated microscope. The herbarium sheets of the relevant taxa have been deposited in the herbaria of Post Graduate Teaching Department of Botany, Nagpur University, Nagpur (M.S). The family does not show one type of embryological character.

OBSERVATIONS

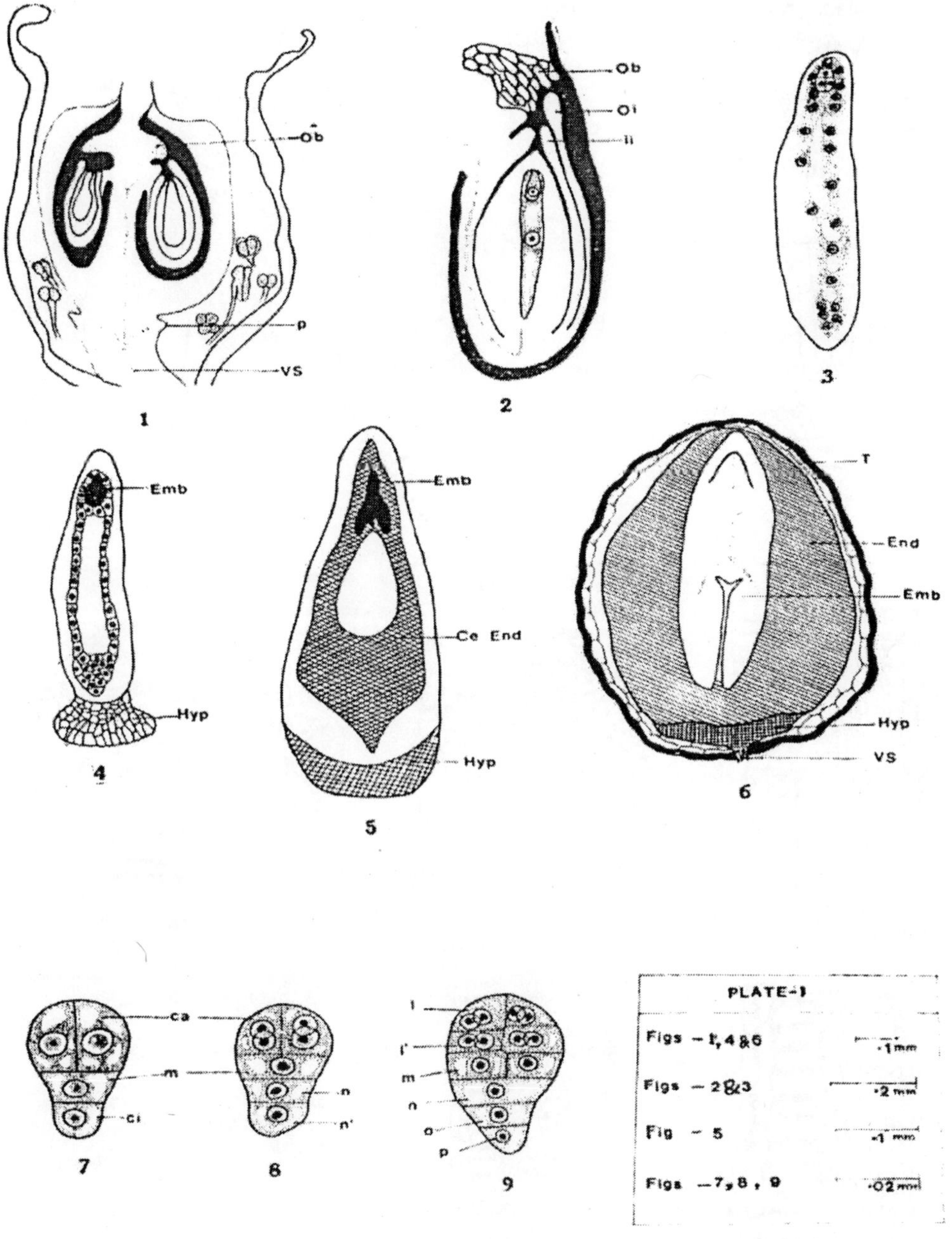

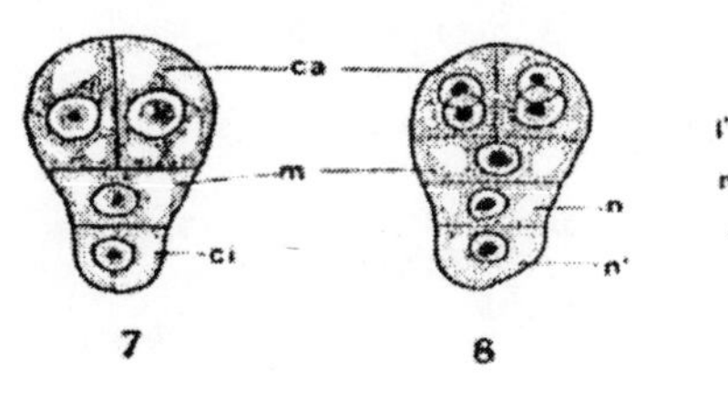

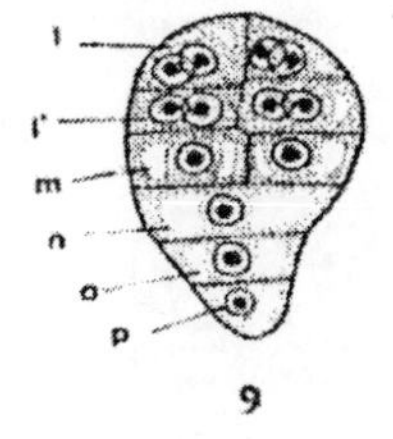

Ob
p
VS
1
Ob
Oi
Ii
2
3
Emb
Hyp
4
Emb
Ce End
Hyp
5
T
End
Emb
Hyp
VS
6
ca
m
ci
7
n
n'
8
I
I'
m
n
o
p
9
PLATE-1
Figs – 1,4&6 ·1mm
Figs – 2&3 ·2 mm
Fig – 5 ·1 mm
Figs – 7,8,9 ·02 mm

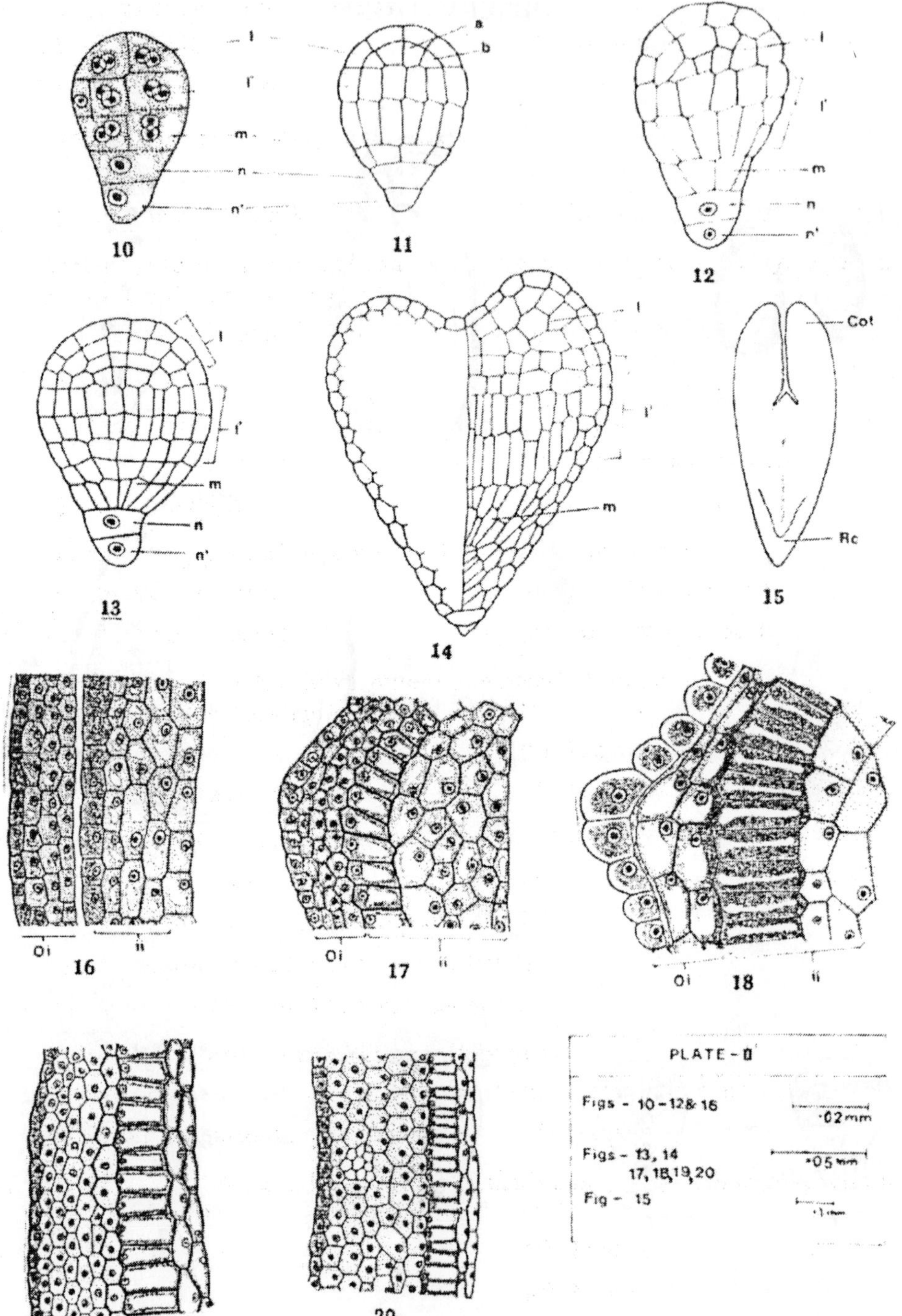

l
l'
m
n
n'
a
b
Cot
Rc
Oi
ii
10
11
12
13
14
15
16
17
18
19
20
PLATE - II
Figs - 10-12& 16
·02 mm
Figs - 13, 14
17, 18, 19, 20
·05 mm
Fig - 15
·1 mm

RESULT AND DISCUSSION

a) Megasporangium:

The centrally placed naked female flower of the cyathium is represented by a superior, tricarpellary, trilocular gynoecium mounted on a long pedicel (fig.1). It is surrounded by five groups of male flowers. As the female flowers develops, it generally bends on one side. Just below the ovary on the pedicel are present distinct articulation or appendage representing the position of vestigial perianth lobe (P). The naked flower in *Euphorbia* should be treated as derived from a complete flower.

Each locule has an anatrapous ovule (fig.1). At the mature embryo sac stage, the ovule becomes bitegemic and crassinucellate (fig.2).

The obturator is an interesting feature in Euphorbiaceae. It is placental in origin and consist of loosely elongated cells. The distal part of the obturator is seen on the exostome. (fig.2).

The nucellus is free from the inner integument in the upper portion (Fig. 2). The micropyle is organized by both the integuments and is almost straight. The vascular supply enters the ovule through the chalaza (fig.2).

b) Endosperm:

The primary endosperm nucleus is situated in the centre of the embryo sac (fig.2). It divides mitotically and at the tetrad stage of embryo a large number of free endosperm nuclei are seen in the embryo sac. (fig.3). Thus the endosperm development is ab-initio nuclear. The endosperm nuclei are at the periphery of the embryo sac in a thin layer of cytoplasm with small central vacuoles. The endosperm nuclei are almost of equal size.

The free nuclear endosperm later on becomes cellular. Cellularization is completed by the time the embryo has become globular (fig.4). Centripetal growth of the endosperm takes place by periclinal division of the first formed cells and finally at the late cordate stage of the embryo, the embryo sac shows cellular endosperm with a small central vacuole (fig.5).

In the mature seed, the embryo sac is completely filled with cellular endosperm (fig.6). At this time, the cells of the endosperm are loaded with reserve food material in the form of aleurone grains. A part of endosperm is utilized by the growing embryo but a considerable part of it persists in the mature seed surrounding the embryo. Thus the seeds are endospermic (fig.6).

c) Hypostase:

Some interesting changes occur in the nucellus after fertilization. Towards the chalazal end, the embryo sac elongates and destroys some of the cells of the nucellus. The remnant of the nucellus at this side, however, become rich in cytoplasm by taking a deep stain and appear more or less regularly arranged. They become conspicuously differentiated to form a thick saucer shaped pad of tissues below the chalazal end of the embryo sac. This is known as hypostase (fig.4,5). The hypostase is a flat saucer shaped structure. No reserve food material was observed in the cells of the hypostase.

d) Embryo:

The zygote gets pronouncedly enlarged before the division take place. It is presumed that the first division is transverse to form cells ca and cb. The next division is the second cell generation where both the cells ca and cb divides, ca divides vertically to form two juxtaposed cells while cell cb divides transversely to form two superposed cells m and ci (fig.7). At the end of this cell generation, the tier ca is of two cells while tier m and ci have one cell each. The characteristic feature of this cell generation is the individualization of the middle cell m which immediately becomes the hypophysis initial and engenders the initials of the root tip and root cap. The pro-embryonic tetrad thus belongs to series A, subseries A2 of Souege's system of classification (see, Crete, 1962).

In the third cell generation, cell ca undergoes one more vertical division at right angles to the first and forms the quadrant Q. At this stage the cell ci divides transversely to form n and n` and m remains undivided (fig.8). Thus at the end of third cell generation, the pro-embryo is of seven cells arranged in tiers. The tier Q has four cells while tiers m, n and n` have one cell each (fig.8).

In the fourth cell generation, all the cells of the quadrant Q divides transversely to form an octant. The eight cells of an octant are arranged in two tiers, designated as 1 and 1` of four cells each (fig.–9). In the meantime n` has divided transversely to form o and p (fig.9). Thus at the end of the fourth cell generation, the pro-embryo is made up of about 12-13 cells disposed in 5-6 tiers. The tiers 1 and 1` consist of 4 cells each while the tier n, o and p consists of one cell each. The tier m if has already divided shows 2 cells (fig.9). The development of the embryo is summarized in a schematic manner below :

I --- Stem tip and cotyledons.

Ca---q

I'---Hypocotyl and radical.

Zygote

m---Hypophysis, root tip and root cap.

Cb

n

ci o

n' Suspensor.

p

It is therefore, evident that the embryo development conforms to Onagrad type of Johansen (1950) and shows *Euphorbia* variation as at the end of third cell generation seven to eight cells are disposed in three to four tiers. According to the periodic system of classification of Souege's (see, Crete, 1962), if falls under grand period I, series A, subseries A2 and megarchetype IV of first group.

Further details regarding the later development of the embryo are as follows :

The terminal tier 1 divides obliquely to form two cells a and b. The inner cell a is rectangular and the outer cell b is triangular (fig.11). In these cells periclinal division takes place to form the outer dermatogens and inner cell (fig.11). The inner cell of a develops into stem tip while cells of b develops into the cotyledons. The cells of the dermatogens divide only in an anticlinal manner keeping pace with the enlarging embryo. In view of the large number of division, taking place in cells a and b the embryo acquires a globular shape (fig.11,12). The meristematic activity in the cotyledonary part is pronounced than the stem tip. This results in the depression (fig.14). This stage in the embryo development is extremely important as it indicates a transient stage from the radial symmetry to bilateral symmetry. The central depression gets very deep while the cotyledons are differentiated as elongated flat blades.

When all these activities are taking place in the tier 1, the tier 1` also divides. The first periclinal division in the pro-embryo take place in the tier 1` to out of the dermatogens (fig.10). The tier 1` divides transversely to form a two tiered structure (fig.11-14). After cutting off the dermatogens the inner cell again divides periclinally to form the periblem and plerome (fig.11,12). The cells of the periblem and plerome both divides transversely and vertically to form a board zone of hypocotyls with the vertical expansion of cells (fig.13,14).

The tier m which was a single cell structure at end of second cell generation divides vertically (fig.9). Another vertical division at right angles to the first take place resulting in four circumaxial cells (fig.10). These cells again divide in periclinal manner. The cells out off on the peripheral side complete the dermatogens in continuation with the dermatogen of 1 and 1` (fig. 11,12). The tier m̀ divides transversely and the inner cells contributes to the formation of root cortex, while the tier nearest to suspensor forms the root tip. (fig.13,14).

The tier n remains undivided but sometimes n` divides transversely to form o and p (fig.9). These tiers form a short suspensor, which is seen up to the cordate stage of the embryo (fig.14). But no suspensor is present in the mature embryo. The mature embryo is straight and dicotyledonous. The cotyledons are flat broad and enclose the stem tip (fig.15). It shows the procambial vascular strand. The length of the hypocotyl is more than the cotyledons. The mature embryo is surrounded by endosperm (fig.6).

e) Seed coat :

The ovules are bitegmic and crassinucellate. During the development of the seed, both the integuments exhibit pronounced changes which are as follows :

(i) Change in the inner integument :

At the embryo sac stage the inner integument is 4 layered (fig.16). The dimension of the inner integument is not uniform. It is broad at the base and narrow at the apex. The cells are polygonal, parenchymatous and compactly arranged (fig.16). At this stage the outer epidermis of the inner integument are darkly stained, while the other layers are vacuolated and show less of staining (fig.16).

At the globular embryo stage, some distinct histological changes take place in the layers of this integument. The cells of the outer epidermis of inner integuments get rapidly distended with thick walls and prominent vacuole (fig.17). Below this radially elongated layer, the sub epidermal layers is seen as enlarged vacuolated cells. In the mature embryo stage the thick walled elongated outer epidermis becomes sclerotic due to deposition of tannin. The sclerotic cells have a narrow lumen. This layer gets adpressed to the inner epidermis of the outer integument. The remaining layers of this integument become devoid of any cytoplasm and are on the way of degeneration (fig.18).

(ii) Changes in the outer integument :

At the mature embryo stage the outer integument comprises of four layers of cells. These cells are more compact and smaller than the cells of the inner integument. The outer epidermis of this integument is darkly stained as compared to other layer (fig.18). Like the inner integument, the outer integument also shows histological changes in conformity with the embryo development.

At the globular stage of the embryo, the seed coat shows definite ridges and furrows which becomes prominent at the mature embryo stage (fig.6). At this stage the seed coat shows a bulge. The outer epidermal cells are darkly stained, the hypodermal layer is tangentially flatten and shows sign of degeneration. The other two layer are highly vacuolated but remain polygonal (fig.17). At the mature embryo stage, the outer epidermal cells are enlarged and a few on the ridges are comparatively more enlarged and bulged. These cells have uniformly thin cytoplasm. By this time the sub epidermal layer has completely degenerated. The other two layers also become enlarged and vacuolated. The inner tangential wall of the inner epidermis become thick walled, perhaps due to deposition of cellulose. The formation of ridges and furrows in the seed coat of outer epidermis of the inner integument during the formation of sclereid layer (fig.18).

(f) Fruit wall :

The ovary wall at mature embryo sac stage consists of 9-10 layers of cells. At this stage the inner epidermis comprises of two layers of tangentially flattened cells. These two layers have originated by the periclinal division of single layer of epidermal cells. The next layer to these 2 layers, become radially elongated and vacuolated. The outer 5-8 layers are compactly parenchymatous (fig.19). At the mature embryo stage, the four inner layers exhibit some histological changes. The two inner epidermal cells become very much tangentially flattened and thick walled. The next layer becomes very much sclerosed and radially elongated followed by another layer of sclerosed cells. The outer epidermis show dark staining contents. The remaining cells are thin walled parenchymatous (fig.20).

From the available data on seed coat structure in Euphorbiaceae it is seen that the number of layers forming the seed coat is variable. The testa is two layered in Peplus (Pal and Khan, 1978), 3 layer in *E. thymifolia* (Pal, 1971), 4 layered in *E. hirta* (Kakale, 1954). 5 layered in *E. cristata* (Mukherjee, 1961 b) and *Croton oblongifolius* (Pal, 1974). Multilayered testa is seen in *Acalypha* (Kapil, 1960 and Mukhejee, 1964) and *Daechampia* (Singh and Pal, 1968).

Though many variations are met with in different embryological aspects within the family, the ultimate structure this taxa the inner epidermis divides periclinally to form 2 layers. Similar behavior is observed in *Chrozophora* (Kapil, 1956) , *Euphorbia hypericifolia* (Mukherjee , 1957), *E. cristata* (Mukherjee, 1961 b) , *E. dulcis* (Kapil, 1961) etc. The hypodermal layer elongates rapidly and is very characteristic of pericarp development. This layer become sclerosed and layer outer to this radially elongated layer becomes thick walled as seen in *Euphorbia* (Mukherjee and Sathianathan, 1985). The remaining layers of the pericarp remain parenchymatous and highly vocuolated.

CONCLUSION

The female naked flower is borne in cup like structure called cyathial cup. This is supposed to be a reduced flower derived from trimerous flower. The tricarpellary, trilocular, syncarpous gynoecium bears anatropous, crassinucellate, bitegmic ovule on an axile placentation. There is single ovule in each locule. The nucellus is enclosed by the integuments. The placental obturator is of loose type. The endosperm is free nuclear, but later on becoming cellular and persist in the seed. The embryo development follows the Euphorbia variation of Onagrad type and falls under Grand period-I, Series A-A_2 and Megachetype IV.

The seed coat develops from both the integuments and the outer epidermis of the inner integument forms the characteristic brittle stony layer of the seed coat. In the pericarp the inner hypodermis is radially elongated and on its both sides the cells are sclerenchymatous.

From the above discussion and conclusion, it is seen that the family does not show one type of embryological characters. The variations are seen in form of ovule, obturator, presence or absence of endosperm haustoria, different types of embryo development and variation in structure and development of seed coat.

Therefore, it is concluded that the family is heterogenous in nature as suggested by Hutchinson (1973).

REFERENCES

1. Bentham, G. and J. D. Hooker. 1883. Genera Plantarum Vol. 3 L. Reeves and Co. Bhanwara, R.K. 1987 Embryology of *E. maddeni* and *E. nivulia*. Curr. Sci. 58: No. 20:1062-1064.
2. Crete, P. 1962. Precis De Botanique. Mason and Co. R'diteurs. Paris.
3. Croniquist, A. 1968. The evolution and classification of living organisms. Mcgraw Hill Co. New York.
4. Cronquist, A. 1982. Synopsis and Classification of living organisms. McGraw Hill Co. New York.
5. Hutchinson, J. 1959 The families of flowering plants. Vol. II 3 Ed. Oxford. Clarendon Press.
6. Hutchinson, J. 1973 The families of flowering plants. 3 Ed. Oxford. Clarendon Press.
7. Johansen, D.A. 1940 Plant Microtechinique McGraw Hill Co. New York.
8. Johansen, D.A. 1950 Plant Embryology. Walthan Mass. U.S.A.
9. Johri, B.M. and R.N. Kapil. 1953. Contribution to the life history of *Acalypha indica* L. Phtyomorphology 3 : 137 – 151.
10. Kajale, L.B. 1954 Fertilization and development of embryo and seed in *Euphorbia hirta* L. Proc. Nat.Inst. Sce. India 4 : 353360.
11. Kapil, R.N. 1956. A further contribution to the morphology and life history of *Chrozophora* Neck. Phytomorphology 8:278 288.
12. Kapil, R.N. 1960. Embryology of Acalypha L. Ibid. 10 : 174 – 184.
13. Kapil, R.N. 1961. Some embryological aspects of *Euphorbia dulcis* L. Ibid. 11: 24-36.
14. Mukherjee, P.K. 1957. Studies in the embryology of *Euphorbia hypericifolie* L. Bull Bot. Soc. Univ. Sagar. 9 :7-18.
15. Mukherjee, P.K. 1961a. Embryology of two Euphorbiaceae Proc. Indian Acad. Sci. 43 : 217- 229.

16. Mukherjee, P.K. 1961b. Study of mature seed coat and fruit wall in *Euphorbia cristata* Heyne. Jour. Bio. Sci. 4: 1-5.
17. Mukherjee, P.K. 1964. Further contribution to the embryology of the genus. Acalypha L. Proc. Nat. Acad. Sci. B. 34 : 129 – 141
18. Mukherjee, P.K. 1965. Contributon to Embryology of *Euphorbia peltata* Roxb. Ibid. 35 : 327- 337.
19. Mukherjee, P.K. and M.D. Padhye . 1964. Contribution to the embryology of the genus Phyllanthus L. Ibid. B 34 : 117 – 128.
20. Mukherjee, P.K. abd K.N. Sathianathan.1985. Embryological studies in *Euphorbia serpens* H.B. K. J. Indian Bot. Soc. Abst. 72.
21. Pal, Aruna. 1971. Structure and development of seed in *Euphorbia thymifolia* Burn. Plant Science 3 : 68 – 71.
22. Pal, Aruna. 1974. Structure and development of seeds in *Croton oblongifolius* Roxb. Acta Botanica, India 2 : 147 – 150.
23. Pal, Aruna. 1984. Seed structure in Hura crepitans L. Nuell – Arg. J. Indian Bot. Soc. Abst. VII – 43 :71.
24. Pal, Aruna., and A.M. Khan. 1978. Structure and development of seeds in *Euphorbia peplus* L. J. India Bot. Soc. Abst. 42 – 43.
25. Pal, Aruna and S. Chopra. 1987. Development and structure of seeds in *Trewia nudiflora* L. Geophytology 617(2) : 241 – 244.
26. Pax, F. And K. Hoffman. 1931. Euphorbiaceae in "Die Naturlichen Pflanzenfamilien" of Engler A. and K. Prantl. 19 C, 11-233. Leipzig.
27. Rao, P.N. 1962. A note on the embryology of *Micrococoa mercurialis*. Benth. Curr. Sci. 31 :426 -427.
28. Rao, P.N., and D.S. Rao. 1975. Embryology of *Phyllanthus rotundifolius* Klein. Proc. Ind. Sci. Cong. 75.
29. Rao, P.N., and D.S. Rao. 1976. Embryology of Cassava (*Manihot esculenta*) Proc. Ind. Nat Sci. Acad. 428 : 111 -116
30. Rendle, A.B. 1925 & 1963. Classification of Flowering Plants. Cambridge Univ. Press. Sathianathan, K.N. and P.K.Mukherjee. 1983. Reproductive ontogeny in two members of Euphorbiaceae. Proc. Ind. Sci. Cong. 61.
31. Singh, R.P. 1962. Forms of ovules in Euphorbiaceae. In Plant Morphology. Syumposium CSIR, New Delhi. 124 – 128.
32. Singh, R.P. 1969. Structure and development of seeds in *Euphorbia helioscopia*. Bot. Mag Tokyo 82 : 287 – 293.

33. Singh, R.P. 1972. Structure and development of seeds in *Phyllanthus niruri* L. J. Indian Bot. Soc. 51 : 73 – 77.

34. Singh, R.P., & A. Pal. 1968. Structure and development of seeds in Euphorbiaceae –*Dalechampia roezeliana* Muell arg. Tech.. Comm Nat. Bot. Gardens. Lucknow 74-85.

35. Singh, S.P. 1959. Structure and development of seeds in *Euphorbia geniculata* Orteg. J. Indian Bot. Soc. 38 : 103 – 108.

36. Venkateshwarlu,J., and P.N.Rao 1963. Endosperm in Euphorbiaceae and occurrence of endosperm haustoria in two species of *Croton*: Curr. Sci.32(11) 514-518.

37. Venkateshwarlu, J., P.N.Rao., and D.S.Rao. 1974. Occurrence of stylar obturator in two Euphorbiaceae. Curr.Sci.23:128-129.

38. Zanwar, C.D. and and P. K. Mukherjee. 1988 Structure and development of seed in *Euphorbia clarkeana*. Proc. 75 Indian Sc. Cong.: Pt.III: Abst.235:164.

39. Zanwar, C.D. and P. K. Mukherjee. 1991. Embryological studies in *Euphorbia coccinea* Roth.Ibid. Abst.220:124.

Floral Diversity and their Conservation (2013), Editors: D.R. Khanna et al.
Published by Biotech Books.
ISBN: 978-81-7622-286-0 Pages: 35-41

4

Providing Health Security through Indian Mulberry Extract

ARCHANA RATKANTHIWAR[1] AND PRAVIN CHARDE[2]

[1] Assistant Professor, Department of Human Development, Home Science Faculty, Sevadal Mahila Mahavidyalaya, Nagpur (M.S.)

[2] Principal, Sevadal Mahila Mahavidyalaya, Nagpur (M.S.)

In the modern way of living, our body faces several negative conditions due to pollution, food habits, stress due to work, financial conditions, living habits, lack of exercise etc. The food we eat does not supply necessary nutrients to cope with the problems our body faces. It is necessary to support the body with adequate nutrients minerals and vitamins to stay healthy. We need to be responsible for our health. Today with the increase in 'on the go' and 'fast food' lifestyle health is ignored to a large extent. It has resulted in an increase in the number of lifestyle related diseases such as heart attack, diabetes, asthma, blood pressure, stress, obesity etc. People are earning money but they are losing health. It also has affected directly the spiritual peace and psychological balance of people. Indian mulberry extract through its amazing divine qualities has come to the rescue of the today's society. It provides health security by working on cellular level. This extract detoxifies the body and maintains hormonal balance. This extract contains more than 150 nutrients which the body need for day to day functioning. With its divine qualities the extract is the best health security provided with no side effect. It ensures the person's wealth at its best level. The researcher studied the disease like blood pressure, diabetes, asthma and mental stress with a sample of 100 patients from 2 clinics of East Nagpur area. It has been observed that after regular intake at least for 3 months it shows positive results in the patient. The extract is providing positive results in various diseases. In today's fast moving and stressful life the extract is very positive and safe for all.

Key words: provides health security, amazing divine qualities, detoxifies the body, no side effects, safe for all.

INTRODUCTION

Indian Mulberry is natural health enhancer. Scientifically this fruit is called, '*Morinda citrifolia*'. The extract of this fruit is having medicinal value, which supports healthy people to stay healthy always and sick people to become healthy. The fruit extract works at cellular level, detoxifies and cleanses the user's body and on the other it builds and strengthens the cells of our body.

Fig. 1: Indian Mulberry Fruits

It is preventive and protective food supplement that works at the cellular level. It is 100% vegetarian and herbal, has no side effects, safe for all and is truly mother nature's gift to humanity. This fruit extract helps to strengthen immune system, relieves pain without side-effects, promotes the self healing mechanism, relieves stress and helps us stay calm and relaxed, improvement in digestion, enhancement in alertness, memory and concentration are observed. It reduces changes of premature onset of age-related diseases such as arthritis, blood pressure, diabetes and mental stress.

How does it work in our body?

The fruit extract works at cellular level. It helps our body's chemical reaction to work better since it has a high energy level. The main chemical reactions happening in our body are –

1. **Synthesis of Protein:** Synthesis includes all the things that our body makes.
2. **Cellular mechanism:** It means the flow of chemically controlled information among the cells, for proper co-ordinated growth of the human body as a whole. Our body is made up of micro-units called cells. Healthy cells lead to healthy tissues, healthy tissues lead to healthy organs; healthy organs lead to healthy systems and healthy systems lead to healthy body. Hence, the health of the cells is essential for a body to be healthy. This Indian Mulberry extract is the best nutritional supplement which works at the root level in order to make the fundamental units healthy and thereby make us live healthier.

AIMS AND OBJECTIVES

1. To study the effect of Indian Mulberry extract on diseases like blood pressure, diabetes, asthma and stress.
2. To study whether extract shows effect within 3 months.

MATERIALS AND METHODS

The methodology pertaining to this study is prescribed as follows:

1. **Selection of area and sample:** A random sample of 100 patients suffering from blood pressure, diabetes, asthma and stress are selected from 2 clinics of East Nagpur area.
2. **Method of data collection:** Two types of tools are used *i.e.* interview of patient before starting the treatment and after 3 months after starting the treatment. Observation method is used to collect data. Mental health checklist test is used to collect data related to stress.

3. **Data collection:** Researcher has contacted the patients in clinics and asked various questions related to their diseases and recorded the answers in questionnaire. The researcher has also observed the check-up process of the doctors as they have used, blood pressure apparatus, glucometer, nebulizer and special questionnaire was prepared to study the stress.
4. **Analysis of Data:** The researcher has utilized her abilities and skills to get required data. After collecting data it is transferred to master chart with the help of this master chart tables and interpretations is done and by using these observations, are done.

RESULTS AND DISCUSSIONS

Table 1: No. of patients selected

S. No.	*Diseases*	*No. of Patients (N=100)*	*Percentage*
1	Diabetes	25	25.0
2	Blood Pressure	25	25.0
3	Asthma	25	25.0
4	Stress	25	25.0
	Total	**100**	**100.0**

25 patients of each disease are selected for observation.

Diabetes: The common symptoms observed in 25 patients suffering from diabetes were profuse urination, profuse thrust, general weakness, vertigo and when blood sugar was checked it was found that sugar was increased. The sugar level before treatment and after treatment is displayed in Table No.2.

Table 2: Patients suffering from Diabetes

S. No.	*No. of Patients*	*Sugar Level*		*Percentage*
1	08	**Before Treatment**	**After Treatment**	
2	08	F – 100-140 Am – 150-200	F – 90-110 Am – 120-140	100%
3	09	F – 140-200 Am – 200-250	F – 90-110 Am – 130-150	91%
Total	**25**	F – 150-250 Am – 250-300	F – 90-110 Am – 130-180	85%

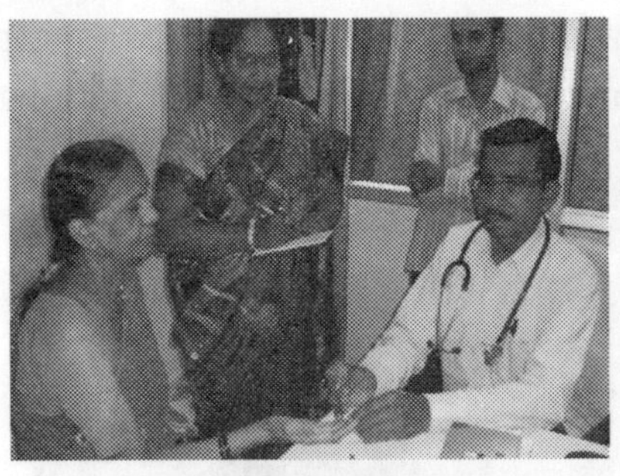
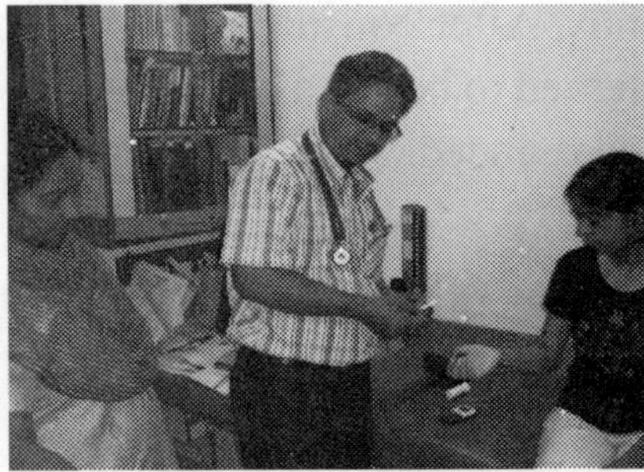

Fig.2: Diabetes Test of Patients by using Glucometer

Blood Pressure: Vertigo, perspiration, general weakness, headache were the common symptoms observed in 25 patients. After consuming I.M.J. for 3 months, all symptoms were found to be in control. Blood Pressure is also in control. Facts are displayed in Table No.2.

Table 3: Patients suffering from Blood Pressure

S. No.	*No. of Patients*	*Blood Pressure*		*Percentage*
		Before Treatment	*After Treatment*	
1	10	D 100 / S 140	D 80 / S 120	100%
2	10	D 120 / S 160	D 80 / S 140	98%
3	05	D 140 / S 200	D 90 / S 150	94%
Total	**25**			

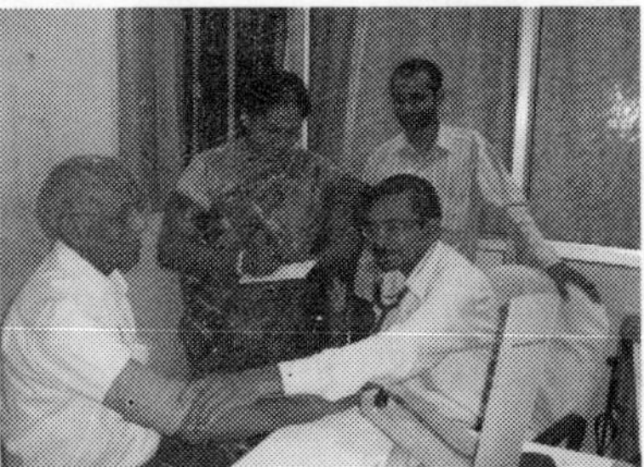

Fig.3: Checking Blood Pressure of Patients by using Blood Pressure Apparatus

Asthma: Breathlessness, frequent cough, weakness, loss of appetite, were the common symptoms observed in patients. Nebulization was given frequently. After consuming I.M.J. continuously for 3 month all symptoms are observed to be control. Frequent cought and use of frequent nebulization has been reduced.

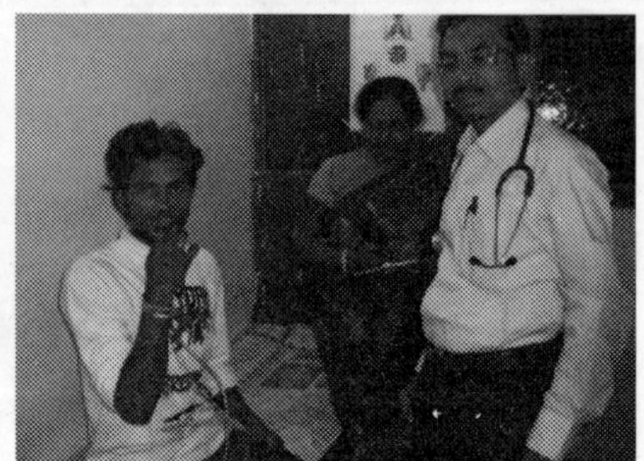

Fig.4: Treatment of Asthma Patients by Nebulizer

Stress: Sleeplessness, depression, anxiety, confusing mind, forgetfulness, lack of concentration and temper tantrum were the common observations found in patients. After treatment all these symptoms are found to be in control.

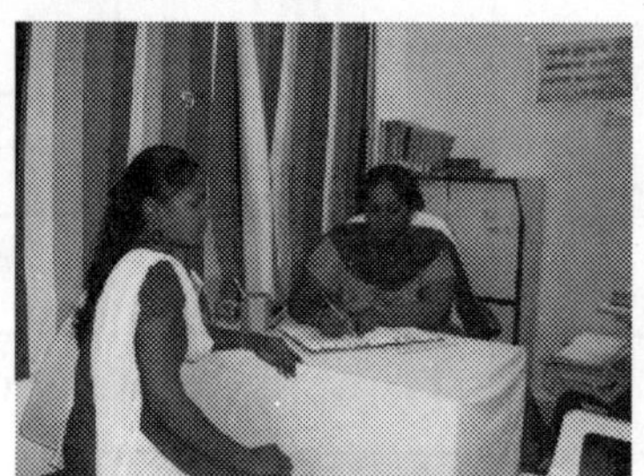

Fig.5: Checking of Stress affected Patients

Positive results are experienced.

For 10% - within 15 days
For 25% - in 1 month
For 50% - in 2 months
For 98% - in 3 months

CONCLUSIONS

1. The researcher has proved that Indian Mulberry extract is safe for all.
2. It can be used as a part of daily diet because of its exceptional nutritional value.

3. It is quality life product.
4. After regular intake of Indian Mulberry Extract, Diabetes, blood pressure, asthma and stress found in control.

ACKNOWLEDGMENTS

Researchers are thankful to Dr. Sanjay Sahare and Dr. Ajay Kale, who are working as general physician in East Nagpur area for providing full co-operation and help in this study.

ABBREVIATIONS

F -Fasting
Am -After meal
D -Diastolic
S -Systolic

REFERENCES

1. Duke, J.A. (1992) Handbook of phytochemicals. Boca Raton, FL: CRC Publishing.
2. Heinicke, R. (2001). The Xeronine system: a new cellular mechanism that explains the health promoting action of NONI and Bromelian. Direct Source Publishing.
3. Levand. O. and Larson H. O. (1979). Some chemical constituents of *Morinda citrifolia.* Planta Med; 36; 186-187.
4. Morton, J. F. (1992). The ocean-going Noni or Indian mulberry (*Morinda citrifolia,* Rubiaceae) and some of its 'colorful' relatives. Economic Botany; 46: 241-256.
5. Merrill, E.D. (1943). Noni (*Morinda citrifolia*) as an edible plant. In: Technical manual: Emergency food plants and poisonous plants of the islands of the pacific. Washington DC: US Government Printing Office.
6. Solomon, N. (1999). The tropical fruit with 101 medicinal uses, NONI juice, 2Ed. Woodland Publishing.
7. www.indian-noni.net

Floral Diversity and their Conservation (2013), Editors: D.R. Khanna et al.
Published by Biotech Books.
ISBN: 978-81-7622-286-0 Pages: 43-49

5

Survey, Collection and Documentation of Brackish Water Algae from East Coast Region of Odisha

H.S. Panda, B.R. Das, B. Parida, J. Jena, P.K. Panda & L.B. Sukla
Bioresources Engineering Department
CSIR-Institute of Minerals and Materials Technology, Bhubaneswar

Odisha has a long coastline of 480 km. extending from West Bengal to Andhra Pradesh. Several rivers, distributaries and channels which drain into the Bay of Bengal, provide a variety of ecological niches and habitats for algal growth. In order to explore the algal biodiversity an extensive field survey was carried out during 2009-2011 and a total of 137 algal samples were collected from 12 sampling sites of various brackish water habitats. About 40 algal strains were isolated and cultured into their pure forms in the Culture Collection Centre & Repository for brackish water algae at CSIR-IMMT, Bhubaneswar. Out of the collected strains 4 taxa of Bacillariophyta, 12 taxa of Cyanobacteria/ Cyanoprokaryota and 24 taxa of Chlorophyta have been identified using relevant monographs and are being screed to evaluate their potential for biofuel production.

Key words: Brackish water, Algae, Bacillariophyta, Taxa

INTRODUCTION

Odisha, located in the east coast of India (Lat. 17° 48'-23° 34' N & Long. 81° 24'-87° 29' E) has an area of 1,55,842 km^2 which is surrounded by West Bengal to the north-east, Jharkhand to the north, Chhattisgarh to the west and north-west and Andhra Pradesh to the south. Several water bodies which drain into the Bay of Bengal, provide a variety of ecological niches and good habitats for the growth of algal species. Though several studies on algal biodiversity assessment in India have been carried out in the past [3, 5], comprehensive survey and collection of algal species in this typical location of the eastern coast region of Odisha as have been done conducted in this study are few [1, 2, 6, 7, 10 & 11]. In the present study field trips were organized to carry out detail survey and collection of algal samples for screening and characterization of potential micro-algae for biodiesel production.

MATERIALS AND METHODS

Algal samples were collected randomly from 12 sampling sites of 8 estuaries of rivers and distributaries from coast of Odisha during 2009-2011 through several collection trips. The estuaries of River Rushikulya a major river at southern part of Odisha is exploited along with its two major distributaries at Bahuda and Gopalpur. On the other hand the estuary of Mahanadi Odishas's largest river at paradeep along with its distributaries Kushabhadra, Nua nai, and Musa nai was exploited for its algal diversity. Devi River was largely occupied by diatoms followed by green algae. Water temperature (26-33°C) and pH (6.0-8.5) was recorded at the time of sampling.

Samples were collected using forceps and needle and/or plankton net (45μm pore size). Epilithic samples were scraped using a tooth brush. All algal samples were put the voucher numbers were preserved in a pre-sterilized specimen bottle with 4% formaldehyde solution. Planktonic samples were fixed with Lugol's Iodine on spot and brought to the laboratory for analysis. Simultaneously the replica of each sample were put in the culture media and brought to the laboratory and incubated. Under microscope, single cells were isolated using glass capillary and put in the culture slants and incubated under light of 3000-4000 lux intensity. Cellular dimension measurement was carried out by micrometry and microphotograph of each specimen was taken using a Meiji ML-TH-05 Trinocular research microscope fitted with

Nikon Coolpix 4500 digital camera. The organisms were identified using relevant monographs for various algal groups [4, 8, 9 & 12]

RESULTS & DISCUSSION

A: Isolation of algal strain

Forty different algal strains were isolated and made into their pure forms and deposited in form of slants in the Culture Collection Centre & Repository for brackish water algae at IMMT (CSIR), Bhubaneswar assigning with a strain number (Table 1).

Table 1: List of algal strains maintained in IMMT Culture Collection / Repository

Sl. No.	*No. of cultures collected so far*	*Strain Number*	*Sl. No.*	*No. of cultures collected so far*	*Strain Number*
1	*Chlorococcum* sp.	IMMTCC-01	**21**	*Oscillatoria* sp.	IMMTCC-21
2	*Chlorella* sp.	IMMTCC-02	**22**	*Oocystis* sp.	IMMTCC-22
3	*Scenedesmus* sp.	IMMTCC-03	**23**	*Chlamydomonas* sp.	IMMTCC-23
4	*Chlorococcum* sp.	IMMTCC-04	**24**	*Scenedesmus* sp.	IMMTCC-24
5	*Chlorella* sp.	IMMTCC-05	**25**	*Scenedesmus* sp.	IMMTCC-25
6	*Scenedesmus* sp.	IMMTCC-06	**26**	*Chlorococcum* sp.	IMMTCC-26
7	*Scenedesmus* sp.	IMMTCC-07	**27**	*Chlorella* sp.	IMMTCC-27
8	*Chlorella* sp.	IMMTCC-08	**28**	*Chroccodiopsis* sp.	IMMTCC-28
9	*Chlorella* sp.	IMMTCC-09	**29**	*Phormidium* sp.	IMMTCC-29
10	*Phormidium* sp.	IMMTCC-10	**30**	*Phormidium* sp.	IMMTCC-30
11	*Chlorococcum* sp.	IMMTCC-11	**31**	*Chlorella* sp.	IMMTCC-31
12	*Chlorococcum* sp.	IMMTCC-12	**32**	*Chlorella* sp.	IMMTCC-32
13	*Scenedesmus* sp.	IMMTCC-13	**33**	*Phormidium* sp.	IMMTCC-33
14	*Oscillatoria* sp.	IMMTCC-14	**34**	*Cymbella*	IMMTCC-34
15	*Bracteococcus minor*	IMMTCC-15	**35**	*Nitzschia* sp.	IMMTCC-35
16	*Chlorella* sp.	IMMTCC-16	**36**	Diatom	IMMTCC-36
17	*Chlorococcum* sp.	IMMTCC-17	**37**	Diatom	IMMTCC-37
18	*Chlorella* sp.	IMMTCC-18	**38**	Nostoc	IMMTCC-38
19	*Phormidium* sp.	IMMTCC-19	**39**	Aphanothece	IMMTCC-39
20	*Phormidium* sp.	IMMTCC-20	**40**	Nostoc	IMMTCC-40

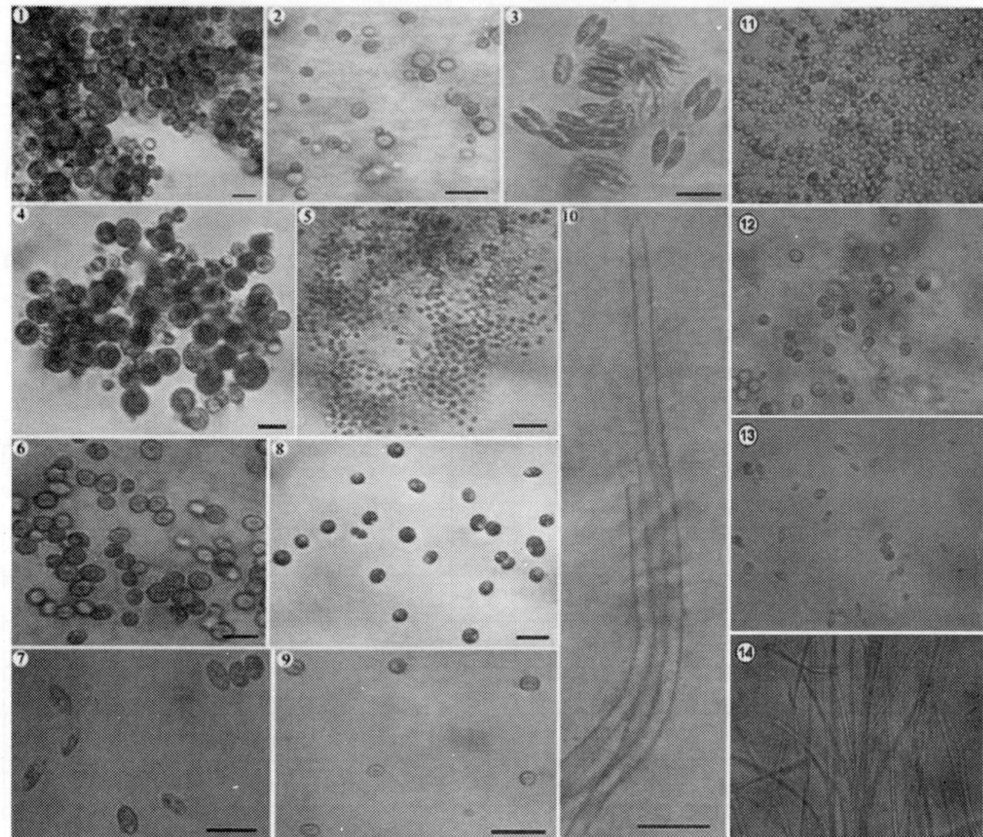

Fig. 1: Algal strains representation of IMMTCC-1, IMMTCC-2, IMMTCC-3, IMMTCC-4, IMMTCC-5, IMMTCC-6, IMMTCC-7, IMMTCC-8, IMMTCC-9, IMMTCC-10, IMMTCC-11, IMMTCC-12, IMMTCC-13 and IMMTCC-14.

B. Documentation of algal taxa

Algal taxa were enumerated morphometrically and documented class wise (Table 2).

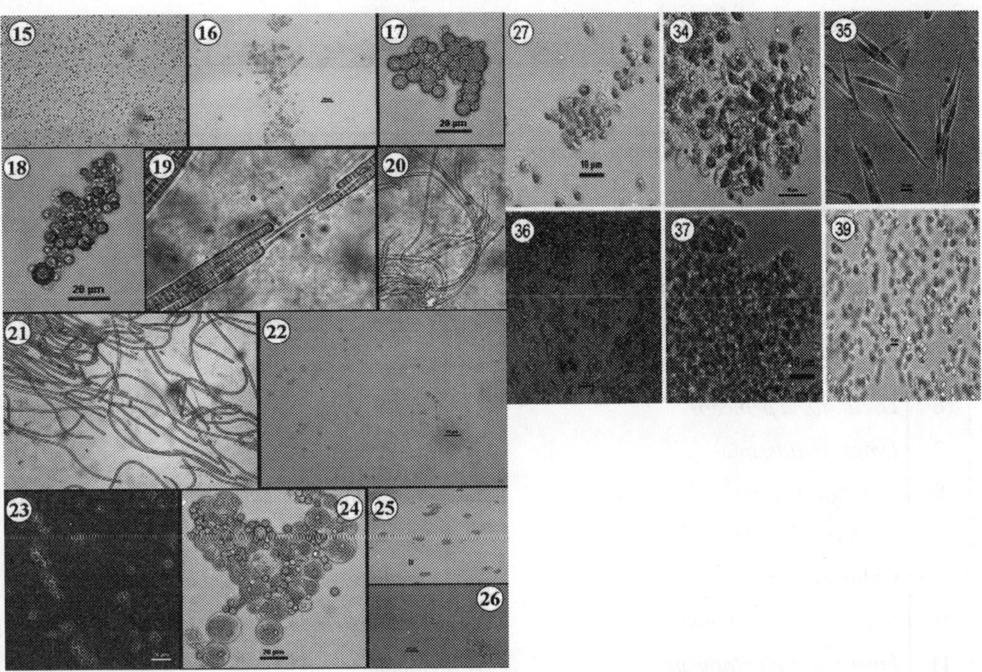

Fig. 2: Algal strains representation of IMMTCC-15, IMMTCC-16, IMMTCC-17, IMMTCC-18, IMMTCC-19, IMMTCC-20, IMMTCC-21, IMMTCC-22, IMMTCC-23, IMMTCC-24, IMMTCC-25, IMMTCC-26, IMMTCC-27, IMMTCC-34, IMMTCC-35, IMMTCC-36, IMMTCC-37and IMMTCC-39.

CONCLUSION

The study reveals that a total of 51 algal taxa out of 137 algal samples have been identified. Out of 51 algal taxa, 9 taxa of Cyanobacteria/ Cyanoprokaryota, 15 taxa of Chlorophyta, 3 Euglenophyta and 24 taxa of Bacillariophyta were reported from these estuaries [Table 2]. Diatoms are found to be dominant in the river estuary followed by green algae and blue green algae. The pure algal strains deposited in our laboratory include 4 taxa of Bacillariophyta, 12 taxa of Cyanobacteria/ Cyanoprokaryota and 24 taxa of Chlorophyta [Table 1]. The green algae deposited in the repository are under process of screening for their potentiality of biofuel production and for further use.

Table 2: List of algal taxa documented from different brackish water habitats from east coast of Odisha.

S. No.	*Organisms*	*S. No.*	*Organisms*
	Cyanoprokaryota/Cyanobacteria	26	*Euglena sanguine*
1	*Microcystis aeruginosa*	27	*Trachelomonas volvocina var. subglobosa*
2	*Chroococcus turgidus*		**Bacillariophyceae**
3	*Planktothrix planktonica*	28	*Cyclotella meneghiana*
4	*Planktothrix compressa*	29	*Cocconeis pediculus*
5	*Phormidium chalybeum*	30	*Gramatophora undulate*
6	*Oscillatoria princeps*	31	*Synedra ulna var. aequalis*
7	*Lyngbya latissima*	32	*Pinnularia microstauron*
8	*Lyngbya holdnei*	33	*Achnanthes coarctata var. parallela*
9	*Anabaena torulosa*	34	*Navicula cuspidate*
	Chlorophyta	35	*Gomphonema lanceolatum*
10	*Stigeoclonium attenuatum*	36	*Gomphonema micropus*
11	*Enteromorpha clathrata*	37	*Gomphonema vibrio*
12	*Pediastrum tetras*	38	*Gyrosigma scalproides var. exima*
13	*Desmodesmus armatus* var. *spinosus*	39	*Nitzschia obtusa*
14	*Scenedesmus acunae*	40	*Nitzschia vasnii*
15	*Scenedesmus acutus* var. *acutus*	41	*Nitzschia longissima*
16	*Scenedesmus dimorphus*	42	*Nitzschia sigma*
17	*Scenedesmus ecornis* var. *ecornis*	43	*Nitzschia closterium*
18	*Scenedesmus quadricauda*	44	*Skeletonema costatum*
19	*Monoraphidium contortum*	45	*Coscinodiscus jonesianum*
20	*Monoraphidium griffithii*	46	*Asterionella japonica*
21	*Monoraphidium minutum*	47	*Achanthoceros zachariasii*
22	*Monoraphidium indicum*	48	*Rhizosolenia alata*
23	*Kirchneriella rotunda*	49	*Staurosirella pinnata*
24	*Coelastrum indicum*	50	*Cyclostephanos dubius*
	Euglenophyta	51	*Diploneis robustus*
25	*Euglena cuneata*		

ACKNOWLEDGEMENT

We thank Department of Bio-technology (DBT), Govt. of India for financial assistance and Prof. B.K. Mishra, Director, Institute of Minerals & Materials Technology (CSIR) for providing laboratory facilities.

REFERENCES

1. Adhikary, S.P. 2000. A preliminary survey of algae of estuaries and coastal areas in Orissa. *Seaweed Res*earch *and Utilization*, 22:1-5.
2. Bhakta, S. Das, S. K. Nayak, M. Jena, J. Panda, P.K. Sukla, L. B. 2011. Phyco-Diversity Assessment of Bahuda River Mouth Areas of East Coast of Odisha, India, *Recent Research in Science and Technology,* 2: 80-89.
3. Biswas, K.P. 1949. Common fresh and brackish water algal flora of India and Burma. *Rec. Bot. Surv. India*, 15 (1): 1-105.
4. Desikachary, T.V. 1959 Cyanophyta. I.C.A.R. monograph on Algae. New Delhi. Pp 686.
5. Habib, I. and Chaturvedi, U.K. 1999. A systematic account of Chlorococcales from Ramnagar, Kumaun Himalaya. *Phykos*, 38: 97-100.
6. Jena, M. Ratha, S.K. and Adhikary, S.P. 2005. Algal diversity changes in Kathjodi river after receiving sewage of Cuttack and its ecological implication. 8: 67-74.
7. Jena, M. Ratha, S.K. and Adhikary, S.P. 2008. Algal diversity of Rushikulya river, Orissa from origin till confluence to the sea. *Indian Hydrobiol.*, 11 (1): 9- 24.
8. Komárek, J. and Anagnostidis, K. 1998 Cyanoprokaryota 1.Teil: Chroococcales. In: Ettl H, Gartner G, Heynig H and Mollenhauer D (eds.), SübWasserflora. Von Mitteleuropa, Gaustav Fischer, **19**: Pp 548.
9. Komárek, J. and Anagnostidis, K. 2005 Cyanoprokaryota 11.Teil: Oscillatoriales. In: Büdel B, Gartner G, Krienitz L and Schagerl M (eds.), SüßWasserflora. Von Mitteleuropa, Elsevier, 19: Pp 759.
10. Nayak, M. *et al.* 2011. Screening of Fresh Water Microalgae from Eastern Region of India for Sustainable Biodiesel Production. *International journal of green energy.* 8: 1–15.
11. Padhi, M. and Padhi, S. 1999. Phytoplanktonic community of Gopalpur estuary. *Seaweed Res. Utiln.*, 21: 95-97.
12. Philipose, M.T. 1967 Chlorococcales. I.C.A.R. Monographs on Algae, New Delhi, Pp: 365.

Floral Diversity and their Conservation (2013), Editors: D.R. Khanna et al.
Published by Biotech Books.
ISBN: 978-81-7622-286-0 Pages: 51-55

6

Icthyofaunal Diversity of Paddy Fields and Streams of Janjgir - Champa District (C.G.)

R. K. TAMBOLI* AND Y. N. JHA**

*Asstt. Professor, Department of zoology, K. Govt. Arts & Science College, Raigarh (C.G.)

**Principal, Govt. Niranjan Kesherwani College, Kota, Bilaspur (C.G.)

India is a tropical country and is bestowed with bountiful natural resources in the form of extensive coastlines, river systems, estuaries, ponds, tanks, lakes, reservoirs and beels. The water bodies are extremely productive and harbour an enviable spectrum of fish genetic resources. There are about 742 freshwater species of fishes under 233 genera, 64 families and 16 orders in India, (Jayaram ; 1981). A number of inland culture fish farming programs are going on but no investigations capturing fishes have been made in the area regarding their icthyofaunal diversity. During the monsoon season these captured fishes are spread in the paddy fields near by. During monsoon and before harvesting the paddy crop, the fishermen, farmers and villagers collect these fishes from the paddy fields and the streams. The people of this area are aware about these fishes as they are economically very important for them. Investigations, identification of these fishes have not been done, thus there is an urgent need for proper investigation and documentation of this fish diversity.

The present study is made to identify the fish fauna of the paddy fields and other related streams, canals of the Janjgir-Champa district. The study was made during the mansoon period, July 2009 to Nov. 2009 and July 2010 to Nov. 2010. During the present studies 31 species of freshwater fishes were identified and they belong to 17 genera, 14 families of class Teleostomi.

Key words : Icthyofaunal diversity, Paddy fields, Sreams

INTRODUCTION

Janjgir - Champa district of Chhattisgarh state has very rich freshwater resources in the form of rivers, irrigation canals, seasonal streams and tanks and these water resources are helpful in natural rearing of the captured fishes.

During the monsoon season these captured fishes are spread in the paddy fields near by. During monsoon and before harvesting the paddy crop, the fishermen, farmers and villagers collect these fishes from the paddy fields and the streams. The people of this area are aware about these fishes as they are economically very important for them. Investigations, identification of these fishes have not been done, thus there is an urgent need for proper investigation and documentation of this fish diversity.

MATERIAL AND METHODS

(i) Collection : The collections of these fishes from the various resources were made during the period from July 2009 to November 2009 and July 2010 to Nov. 2010. The collections were specially made with the help of local fishermen, farmers and from the local fish market and were brought to the zoology department for identification and preservation.

(ii) Identification : The fishes so collected were identified with the help of fin formula and scale counts as per Day' 1878, Menon, 1987, Jayaram, 1991 and Talwar and Jhingran, 1992.

(iii) Preservation : The fishes were eviscerated and preserved in 10% formalin.

RESULT AND DISCUSSION

During the course of studies about 32 species of fishes were recorded and they belong to 17 genera and 14 families. The wide spread species belong to the families, Anabantidae, Bagridae, Belonidae, Centropomidae, Channidae, Cichlidae, Claridae, Cyprinidae,Gobiidae, Heteropneustidae, Mastacembelidae, Nandidae and Siluridae and the rare species of Family Amphipnoidae. The most common varieties are *Anabas*, *Channa*, *Clarias*, *Chanda*, *Esmos*, *Mystus*, *Heteropneustus*, *Mastacembelus*, *Puntius*, *Tilapia* and *Xenentodon* species. The *Wallago pabda* and *Nandus nandus* are common but *Amphipnous cuchia* is very rare and occasionally seen. This species is medicinally important and sold at high price.

Table 1: List of fishes of paddy fields and streams of Janjgir - Champa district

S. No.	***Local Name***	***Scientific Name***	***Family***	***Source***
1	Rukh chagha	*Anabas testudineus*	Anabantidae	Paddy fields
2	Malaj baam	*Amphipnous cuchia*	Amphipnoidae	Streams
3	Baril	*Barilius baril*	Cyprinidae	Streams
4	Chandni	*Chanda nama*	Centropomidae	Paddy fields, Streams
5	Mongri	*Chanda ranga*	Centropomidae	Paddy fields, Streams
6	Mrigal	*Cirrhinus mrigala*	Cyprinidae	Paddy fields
7	Karajiya	*Channa gachua*	Channidae	Paddy fields, Streams
8	Khoksi	*Channa marulius*	Channidae	Streams
9	Khoksi	*Channa punctatus*	Channidae	Paddy fields, Streams
10	Bhunda	*Channa striatus*	Channidae	Streams
11	Kewai	*Clarias batrachus*	Claridae	Paddy fields, Streams
12	Magur	*Clarias gariepinus*	Claridae	Streams
13	Dadia	*Esmos danricus*	Cyprinidae	Paddy fields, Streams

contd...

S. No.	Local Name	Scientific Name	Family	Source
14	Ghesra	*Glossogobius giuris*	Gobiidae	Paddy fields Streams
15	Singhi	*Heteropneustus fossilis*	Heteropneustidae	Paddy fields, Streams
16	Rohu	*Labeo rohita*	Cyprinidae	Paddy fields, Streams
17	Bata	*Labeo bata*	Cyprinidae	Streams
18	Calbasu	*Labeo calbasu*	Cyprinidae	Streams
19	Gonius	*Labeo gonius*	Cyprinidae	Streams
20	Tengna	*Mystus bleekeri*	Bagridae	Paddy fields, Streams
21	Tengna	*Mystus cavasius*	Bagridae	Streams
22	Tengna	*Mystus Tengara*	Bagridae	Paddy fields, Streams
23	Tengna	*Mystus Vittatus*	Bagridae	Paddy fields, Streams
24	Baam	*Macrognathus aculeatus*	Mastacembelidae	Paddy fields, Streams
25	Baam	*Mastacembelus armatus*	Mastacembelidae	Paddy fields, Streams
26	Baam	*Mastacembelus Puncalus*	Mastacembelidae	Paddy fields, Streams
27	Bhedo	*Nandus nandus*	Nandidae	Paddy fields, Streams
28	Sarna	*Puntius sarana*	Cyprinidae	Paddy fields, Streams
29	Kotri	*Puntius ticto*	Cyprinidae	Paddy fields, Streams
30	Telpia	*Tilapia mossambica*	Cichlidae	Paddy fields, Streams
31	Pabda	*Wallago pabda*	Siluridae	Streams
32	Sodhia	*Xenentodon cancila*	Belonoidae	Streams

ACKNOWLEDGEMENTS

The author is thankful to Dr. A. Hundet, Head, Department of Zoology and to Dr.A. K. Mehar, Principal, K. Govt. Arts & Science College, Raigarh for providing the department for research work. The author is also grateful to Mr. P. P. Nande Asstt. Director Fisheries, for his valuable suggestions and in the identification of the fishes. Help given by Fishermen and farmers of Janjgir - Champa district in sampling is gratefully acknowledged.

REFERENCES

1. Bakawale, S. and R. R. Kanhere (2006) Fish fauna of river Narmada in West Nimar (M.P.) Research Hunt., 146-51
2. Day, F., 1878, Fishes of India (vol. I and II)
3. Heda, N. K. (2009). Freshwater Fishes of Central India: A Field Guide.(2009). Vigyan Prasar, Department of Science and Technology, Government of India, Noida,169 pp.
4. Jayaram, K. C., 1981, Fresh water fishes of India. Hand book, Zoological Survey of India. Calcutta.
5. Jayaram, K. C., 1999, The freshwater fishes of the Indian region. Narendra Publishing House, New Delhi.
6. Menon, A. G. K., 1987, Fishes of India and adjacent countries (Pisces)vol. 4 (Part I) ZSI, Calcutta.
7. Shrivastava, G.J., 1968, Fishes of Eastern Uttar Pradesh, Vishwavidyalaya Prakashan, Varanasi
8. Talwar, P. K., and Jhingran, K. C., 1991, Inland fishes of India and adjacent countries.(vol. I and II) Oxford and IBH Publishing Co. Pvt. Ltd. New Delhi
9. Tilak, R. and Tiwari, D. N.,1976, On the fish fauna of Poona district (Maharashtra) News letter. ZSI, 2 (5): 193 -199.
10. Verma, K. P., 1977, Fishes of North- Western (Tirhut division) Bihar, thesis accepted for Ph. D. degree, University of Bihar, Muzaffarpur.

Floral Diversity and their Conservation (2013), Editors: D.R. Khanna et al.
Published by Biotech Books.
ISBN: 978-81-7622-286-0 Pages: 57-68

7

Biodiversity and Food Resources: The Man-nature Relation from Indigenous Cognition

SOMA BANDYOPADHYAY
Assistant Professor and Head Post graduate Department of Anthropology
Sree Chaitanya College, Habra, West Bengal

Food is a substance tinged with emotion from its associations with early care and is a focus of beliefs and myths regarding its properties. Various relationships can be symbolized through food and a range of emotions can be expressed in concrete form. The ways of obtaining, preparing, presenting and sharing are determined by cultural prescriptions and emotional associations as well as by the physiological need to eat. Choice and definition of food from a wide range of available plant and animal resources differentiate people from their non human competitors. A person's relationship with nature is fashioned through individual and general, temporary and permanent proscriptions of certain foods. Certain prescription or proscription is undoubtedly regulated by in-depth knowledge about the biodiversity of the area. Perfect management of food resource is only possible through proper consumption-conservation equilibrium which is thoroughly controlled by knowledge about source i.e environment and its variation i.e bio-diversity. The World Food Summit of 1996 defined food security as existing "when all people at all times have access to sufficient, safe, nutritious food to maintain a healthy and active life". The quality parameters of food like safe, nutritious etc are culturally defined. This article deals with this cultural definition of food cognized by an indigenous group of people residing within biodiversity of forest in Jharkhand, India. The **Kol** cognition of food and its relation with local biodiversity has been analysed here with special reference to their consciousness about food security and maintenance of environment.

Key words: Food, Emotion, Cultural prescription, Knowledge, Biodiversity, Management of food resource, Consumption-conservation equilibrium, Food security, Indigenous cognition.

INTRODUCTION

Analytical thinking about the natural world is a key phenomenon of human culture. Rational analysis results into in-depth perception about the natural objects surrounding him. The degree of perception has a close relation with utility. A natural object having the optimum utility value to maintain human existence becomes resource for the person or community utilizing the same. The way of utilization has a keen relation with the unique crown of man – his culture. Culture of a particular community initiates the perception of grading the natural objects in qualitative point of view (Bandyopadhyay, 2007). Logic and wisdom of the community members give this a definite shape. Cognitive anthropology argues that total sphere of human thought is culturally significant. Communicable cognitive features are more significant as human thought is documented within those features. Language is one of the most dependable media for transmission of knowledge as well as documentation of thought.

METHODS

Through continuous conversation with the folk people, different aspects of their "cognized environment" are explored in this study. This exploration highlights the standardized mental makeup, which the community members possess. A universal feature is found at least so far as their common sense regarding a specific natural or social aspect is concerned. The Universal feature is consensually distributed among the folk people. So the study can be considered undoubtedly as a study of *cultural consensus*. Ethnic point of view regarding different perceptual factors has given utmost priority. The study has been done with the methods prescribed within the discourse of ethnoecology and cognitive anthropology. Along with it ethnoscientific orientation of different facts of life is also recorded. Regarding this orientation folk logic and wisdom get priority importance and the knowledge that do not sustain peoples' logic and wisdom is exempted from recording.

FOOD AND NATURE

Gilmore (1932: 320) emphasized the fact that tribes such as Indian tribes depended immediately upon their floral environment first of all for food. Food has always been more than something to steal hunger (de Garine, 1972: 143). Food is a substance tinged with emotion from its associations with early care and is a focus of beliefs and myths regarding its properties (Frazer, 1922: 556). Folk perception and cultural practices related to food supplies lot of information regarding their cultural activities. Food is something more than simply to eat. Reid (1986: 49) explains food as a medium of exchange. According to Reid various relationships can be symbolized through food and a range of emotions can be expressed in concrete form. The ways of obtaining, preparing, presenting and sharing are determined by cultural prescriptions and emotional associations as well as by the physiological need to eat (Richards, 1932; Bennett, 1943; Simoons, 1961; Young, 1971; de Garine, 1972; Schieffelin, 1977). Manderson, (1986a: 6) states that "Food serves to distinguish humanness. Choice and definition of food from a wide range of available plant and animal resources, the preparation of selected and acceptable resources prior to consumption, and ideas regarding the appropriate consumption of foods according to time of a day and in association with other foods differentiate people from their potential and actual non human competitors for food".

In this study cultural distinctiveness of the **Kol** residing in east Singbhum District of Jharkhand, India has been dealt with.

While studying among the **Semai** of Malay peninsula, Dentan (1968: 45) highlighted the inextricable link between humans and food. Each and every culture has specific prescription regarding cultivation of staple food product. Similar to **Kalauna** people studied by Young (1971: 195-198) the **Kol** distinguish their humanness establishing their ability to produce staple crops or ***Baba*** (paddy). Another distinctive feature of human beings regarding food is specificity in selection of edible and inedible varieties. A person's relationship to his or her environment is fashioned through individual and general, temporary and permanent proscriptions of certain foods (Manderson, 1986a: 8).

Consumption of food is designated by the **Kol** as ***Jomlekhan.*** They have specific terminology for solid food - ***Jom***, but specific terminology for food as a whole is not so prevalent. Dichotomy of edibility is represented as ***Jomkida*** (edible) and ***Bakojomkida*** (inedible) respectively. Habitat and nature of the food item and its relation to human beings are the basis of inclusion in the list of prescribed food. According to frequency of consumption food items are classified into three groups. ***Asol jom*** that means staple food, ***Dimsi Jom*** (accompanied food items regularly consumed) and ***Musing Jom*** (accompanied food items occasionally consumed). Mainly non-flesh foods like ***Ara*** (leafy vegetables) and ***Utu*** (green vegetables) are cognized as ***Dimsi Jom*** and fleshy foods are included within the ***Musing Jom*** category. ***Bulung*** (salt) is a common item in both the categories. Conditions of intake gives rise to another set of categories. The raw uncooked foods are categorized as ***Berel te*** and the cooked foods are considered as ***Ising kate*** as the word ***Ising*** means cooking.

COGNITION OF PRINCIPAL FOOD

Mandi or rice is the principal food. Rice is consumed in four cuisines namely –***Basi mandi*** (soaked rice), ***Aiyer mandi*** (gruel rice), ***Lolo mandi*** (cooked warm rice) and ***Letto mandi*** (medley rice). Prescription or prohibition of a specific cuisine is highly influenced by the ethnoscientific knowledge. ***Basimandi*** (soaked rice) is the most frequently consumed cuisine of rice for a normal adult with perfect health condition but a restricted one for ***Rua kanah*** (a sick person). A person suffering from ***Ruhe*** (fever) or ***Thetna husui*** (pain in knee), will not be allowed to consume soaked rice as ***Basimandi*** is ***Esu dah kana*** (contains more water). ***Poati*** (pregnant) woman should strictly avoid soaked rice because ***Basimandi joma hormor riyarah*** (intake of soaked rice keep the body cool) and this feature may affect the growing baby. ***Lolomandi*** (cooked warm rice) is used as a symbol of relationship as it is occasionally consumed in order to entertain ***Ket koya*** (guest) or ***Kutum*** (relative). Season also has a lot to do with peoples' preference towards specific type of rice. People like warm ***Aiyermandi*** (gruel rice) in ***Rabang din*** (winter season) and ***Lettomandi*** (medley rice) in ***Dah din*** (rainy season).

COGNITION OF ACCOMPANIED FOOD

Mainly ***Ara*** (leafy vegetables) and ***Utu*** (green vegetables) are consumed as accompanied food but no specific term is used. Here availability gets priority over preference. ***Ara*** are those leafy vegetables, which are mostly cultivated in ***Bakai*** (kitchen garden). People can name a few, maintaining the order of hierarchy in respect to frequency of consumption. ***Lepera*** (*Amaranthus gangeticus*) is the most common cultivated leafy vegetable, followed by ***Mulga*** (*Moringa pterygosperma*), ***Puroi*** (*Basella rubra*), ***Pundi*** (*Vigna catjang*) and ***Muleya ara*** (leaf of *Raphanus sativus*) in hierarchical order.

Categorization of ***Utu*** (green vegetable) is done according to the place of their growth. Vegetables grown underground are identified as ***Hasa bhitire utu***. The other category is ***Hasa chetan utu*** or vegetables growing upon the earth surface.

Some food items are declared as prohibited, depending upon different factors like season, habitat, specific health condition and some others. ***Mulge ara*** (*Moringa pterygosperma* leaf) is not preferred in ***Dah din*** (rainy season), as the leaves contain different insects in this period. ***Alti*** (*Alocasia indica*) if grown aside a pond is not consumed as these ***Alti*** absorb more water and the consumer feel ***Babat*** (itchy feeling) in throat. Children are not allowed to consume ***Sirgiti ara*** (*Chenopodium album*) as it may cause ***dah ge*** (diarrohea). Aged persons avoid ***Piyaj*** (*Allium cepa*), ***Ada*** (*Zinjiber officinale*) as these are considered as ***Hadh jom*** (rich food) and may disturb digestion. Accompanied foods are also classified within the dichotomous domain of flesh and non-flesh food. People distinctly include floral consumable resources within ***Ka jilu jom*** meaning food without flesh and faunal consumable resources within ***Jilu jom*** meaning food with flesh.

Etymologically ***Jilu*** means animal flesh and people use name of the species as a prefix while reporting about different edible varieties. For example they report about ***Magri jilu*** (flesh of *Clarias batrachus*), ***Merom jilu*** (flesh of *Capra* sp.) etc. But people include ***Jarom*** (egg) also within the domain of ***Jilu jom***. It may be due to peoples' cognition about animal protein rich food for which specific vocabulary (***Jilu jom***) is available.

Among the **Hawaiians** it was found that people categorized food into two principal categories. ***Ai*** (starch food) ***Ia*** (accompanied food). The term ***Ia*** covered three categories, fish or any marine animal, meat or any flesh food and any food eaten as a relish with the staple including meat, fish, vegetable or even salt. (Pukui and Elbert, 1965; Titcomb, 1967:21). ***Ia*** includes all other food items except for staple starch food. But in case of the Austric speaking **Kol** community this kind of inclusion is hardly found. They use separate term for vegetables, salt etc. and include fish, meat within the domain of ***Jilu jom***. This kind of specification and linguistic coding may be considered as an indicator of accurate perception. Frequent consumption of such kind of food items depicts close and distinct man-nature relationship. So it can be inferred that though etymological similarity is there, **Kol** people have a clear perception about flesh food and to be more precise the **Kol** people categorize non-vegetarian items as a whole within the domain of ***Jilu jom***.

The **Kol** consume six types of ***Jilu*** and hierarchically categorize them according to frequency of consumption. The hierarchy is justified with different logical deductions like availability, financial condition etc. The cultural factors are more prominently expressed in taste perception. ***Merom jilu*** is given topmost priority for its cost effective usage. ***Merom*** (*Capra* sp.) produces more amount of flesh which is profitable for bulk consumption. Consumption of meat and its relationship with social status is another important criterion of choice. ***Ket koya*** (relative) are entertained with ***Sim jilu*** (chicken) in simple occasion if they like to have. But if they come to join the occasion of marriage or village feast, ***Merom jilu*** will be served to maintain a social status.

Fish is also included within the domain of occasional accompanied food. Almost all types of fishes available within the territory are consumed by the people. The **Kol** people can easily fashion a hierarchical arrangement of the fish varieties depending upon availability as well as frequency of consumption. This classification depicts the fact that availability regulates the thorough preference and perception regarding consumption.

Presumable effects created by different fish varieties are also perceived by the people supported by their own logic. The first thing people perceive about ***Icha hai*** (*Penaeus indicus*) is the ***Babat*** (itchy) feeling, which some of them came across whenever they consumed it. They clearly explain that ***Iche hai jom to joto hor bang phabawa***, meaning all people can not consume the ***Iche hai***. Consumption of ***Boar*** (*Wallago attu*) also results the same effect. People logically establish the benevolent effect of ***Kulcha hai*** on their body. They state that if a person suffers from ***Etang mayam hasui kanah*** (dilution of blood), he / she should consume ***Kulcha hai*** regularly.

Some prohibitions are also there. Pregnant women are never allowed to consume ***Rerant hai*** (*Mystus* sp.) as it is believed that consumption of ***Rerant hai*** disturbs downward movement of the growing baby within mother's womb. This belief has a lot to do with the nature of the fish. They argue that ***Rerant hai*** has a nature of coming upward in the water surface. Similarly consumption of the same may affect the movement of the growing baby.

HOT AND COLD FOOD

People everywhere commonly conceptualize certain basic distinctions of food in terms of hot and cold effect on body. Jensen (1966) found that aboriginal people of Malayo-Indonesian region use a system of hot and cold bath metaphorically in response to physiological changes. Manderson (1986b: 127-28) explains this hot – cold dichotomous situation clearly as it is not related to temperature but to its reputed effect on the body. In Malayan English terminology, hot food is said to be "heaty" and a cold food has a "cooling effect" (Tongue, 1974: 77-78).

Riya jom (cold food) includes almost all fruits, vegetables and country liquor and but specific vegetables are incorporated within the ***Hadhadath jom*** (hot food) category. Laderman (1981) found a different story among the Malay aboriginals. In that case alcohol and herbal preparations are perceived as hot. Vegetable like ***Piyaj*** (*Allium cepa*) is perceived in binary opposition according to condition of consumption and effect on health. People are of opinion that ***Berel piyaj*** keep them cool and it is healthy to consume it frequently during ***Situm din*** (summer). Almost all kinds of fruit are cognized as cold food, but specific effect of specific fruit variety is distinctively mentioned. Beneficial effects of different fruit varieties are perceived by the people, in different ways. Regarding ***Piyari*** they describe ***Piyari jom lekhan hurnur dah pare jaya***, which means consumption of ***Piyari*** (*Pisidium guajava*) meet up the need of water in body. ***Kanthar jom lekhan riyare dohoya***, meaning consumption of ***Kanthar*** (*Artocarpus* *heterophyllus*) keep the body cool. ***Kaira jom lekhan hormor mayam pare jaye***, meaning consumption of ***Kaira*** (*Musa paradisica*) maintains blood content in body. ***Kunde bili jom lekhan dah sen chabi aye*** meaning consumption of fruit of ***Kunde*** (*Rubus fruiticosus*) keep the bowel clear. These kinds of logical explanations undoubtedly resulted from prolonged experimentation and cultural prescription leading to concrete cognition.

Levi Strauss (1969) argued that the classification of hot and cold can perhaps also be linked with the opposition of nature and culture. Manderson (1986b : 143) suggested that cold food is generally raw and is closer to nature; hot foods, through cooking are incorporated into culture. But this argument was not thoroughly applicable in this case. In gross analysis it may be considered that as because fruits are the natural objects and can be consumed in raw condition, people perceive them as ***Riya*** (cold) so, Manderson's argument is quite acceptable. But sometimes some cooked food like ***Ul utu*** (sour dish produced by *Mangifera indica*), ***Berel kaira utu*** (curry of *Musa sapientum*), ***Hotor utu*** (curry of *Lagenaria vulgaris*) are consumed to keep the body cool. On the contrary food variety like ***Rasun*** (*Allium sativum*) which they sometime consume in raw condition is perceived as ***Hadhadath*** (hot). Imbalanced consumption of fruit of ***Keora*** (*Holarrhena antidysentrica*) may cause some hot effect on body it is often consumed in raw condition to prevent dysentery.

CONCEPT OF NUTRITION

Awareness regarding the nutritive aspect of food stuff is praiseworthy. Nutrition is cognized through its holistic effect on body and expressed through a phrase - ***Noako jom lekhan hormor bhagiya*** meaning nutritive food items are those, which build up the body. A strong ***Keteyeit*** (musculature) is considered as the outcome of consuming ***Esu bhugin jom*** (good food regarding nutrition). Benign foodstuff, not creating any adverse effect on health is perceived as ***Saba jom*** (normal food). ***Asol jom*** (staple food), is considered as ***Saba*** among which ***Mandi*** (rice) gets the highest priority. Similar case was witnessed by Laderman (1981), among the **Malay** aboriginals. They consider rice and fresh water fish as normal food. Universally the cultural prescription of any human group selects those items as their staple food, which never cause any health hazard.

CONCEPT OF LIQUID FOOD

The dichotomy solid and liquid food is distinctively found in Kol cultural prescription perceived by the folk community. Liquid food is perceived as ***Ynu***. The informants mention mainly four liquid items - ***Dah*** (water), ***Toa*** (milk), ***Pahrua*** (country liquor) and ***Dieng*** (another variety of country liquor). ***Cha*** (tea) is occasionally consumed and naturally not considered as a regular ***Ynu***. ***Toa*** (milk) is collected from different domesticated species like, ***Merom*** (goat), ***Dangri*** (cow) and ***Kara*** (buffalo). Adult persons normally never consume ***Toa***. A ***Hasui kanah*** (sick) person is allowed to consume ***Toa*** without considering the age. ***Toa ynu lekhan hormor dare aguiya*** - consumption of milk resumes energy. Energy gain is hierarchically explained. The members place ***Bitkil toa*** (buffalo milk) on the topmost position indicating maximum yield of energy and ***Merom toa*** (goat milk) at the bottom. Frequency of consumption is directly proportional to availability and accordingly ***Merom toa*** gets the topmost priority.

Rice-beer is termed as ***Dieng***. This is produced in almost all households. Difference in intensity of ***Nisha*** (intoxicating effect) results in differential perception about liquor varieties. ***Pharua***, produced from ***Mohul*** (*Bassila latifolia*) yield more ***Nisha*** (intoxication) and the consumer can eat other diets normally. In ***Dieng***, intensity of intoxication is low and also disturbs normal physical condition especially normal feeling of hunger. Still they prefer to drink ***Dieng*** for valid economic reason as well as it can be easily prepared from ***Souia mandi*** (fermented rice) in home, but production of ***Pharua*** needs a lot of ingredients and technical effort.

CONCLUSION

The World Food Summit of 1996 declared that food security exists when all people at all times have access to sufficient, safe, nutritious food to maintain a healthy and active life. Food security happens when all people at all times have access to enough food that...

- is affordable, safe and healthy
- is culturally acceptable
- meets specific dietary needs
- is obtained in a dignified manner
- is produced in ways that are environmentally sound and socially just

Food Sovereignty: A Right for All - a Political Statement of the NGO/ CSO Forum for Food Sovereignty (13 June 2002, Rome) established food sovereignty as

- is the right of peoples and communities to safe, nutritious, culturally appropriate food, to food-producing resources, and to the ability to sustain themselves.
- ...is the right of peoples and communities to define their own agricultural, labour, fishing, food and land policies which are ecologically, socially, economically and culturally appropriate to their unique circumstances.

It is time now to respect the cultural significance of food habit of a specific community. What particular type of food is nutritious and what is not, which builds up a healthy body and which does not, which cuisine keeps the body cool and which one creates inconvenience is consensually accepted and the optimization comes through prolonged experimentation with trial and error. Social dignity is merged with consumption of a specific food item, which is prescribed by the culture of a community. Food *sovereignty* in true sense can be established only when clear and in-depth knowledge of bio-diversity as well as natural resources and their utilization pattern of a community to maintain consumption-conservation equilibrium is considered with proper dignity. In this discourse knowledge about bio-diversity of **Kol** community and cultural significance of different life-events regarding food has been discussed. Indigenous knowledge considers food not only as an object basically needed for physical existence, but as an object projecting and protecting the identity of a culture too. Consumption of a particular food item has a lot to do with affordability, availability etc. but even in a crisis situation people are not ready to accept an affordable or available item which is culturally prohibited. People are ready to be indebted to entertain their relatives with ***Merom jilu*** but will not arrange the cheaper ***Sukri jilu*** which will stand for the cultural prescription of maintaining social dignity. Intense analysis of the bio-diversity makes the people expert to judge the qualitative aspect of each and every food item. Hierarchical arrangement of food items changes according to the aspects of quality and frequency of consumptions. This study reflects the fact that man-nature relationship is not only restricted within the domain of utilization but it has got immense cultural significance too. The **Kol** cognition of food establishes the fact that awareness about bio-diversity comes only through prolonged co-existence of man and nature which is the one and only basis of human existence in a global village.

REFERENCES

1. Bandyopadhyay Soma 2007: *Human Cognition in Management of Natural resources and Social Relations*. Sarup & Sons, New Delhi.
2. Bennett. J. W. 1943: Food and social status in a rural society. *American Sociological Review* 8: 561-569.
3. Dentan R. K 1968: *The Semai: a nonviolent people of Malaya*. Holt, Rinehart and Winston: New York.
4. de Garine I 1972: The sociocultural aspect of nutrition. *Ecology of food and nutrition* 1: 143-163.
5. Frazer J.G 1922: *The golden bough: a study in magic and religion*. Macmillan Press: London.
6. Gilmore M 1932: Importance of ethnobiological investigation. *American Anthropologist*. 34: 320-327.
7. Jensen E 1966: Iban birth. *Folk* 8-9: 165-178.
8. Laderman C 1981: Symbolic and empirical reality: a new approach to the analysis of food avoidance. *American Ethnologist* 8(3): 468-493.
9. Levi-Strauss C 1966: *The savage mind*. University of Chicago Press: Chicago.
10. Manderson L 1986(a): Introduction: the anthropology of food in Oceania and southeast Asia. In *Shared wealth and symbol: food, culture and society in Oceania and southeast Asia*. Ed. by L. Manderson. Cambridge University Press: New York. Pp: 1-25.
11. Manderson L 1986(b): Food classification and restriction in peninsular Malaysia: nature, culture, hot or cold? In *Shared wealth and symbol: food, culture and society in Oceania and southeast Asia*. Ed. by L. Manderson. Cambridge University Press: New York. Pp: 127-143.
12. Pukui M and S. Elbert 1965: *Hawaiian English dictionary*. Hawaii.
13. Reid J 1986: Land of milk and honey: the changing meaning of food to an Australian aboriginal community. In *Shared wealth and symbol: food, culture and society in Oceania and southeast Asia*. Ed. by L. Manderson. Cambridge University Press: New York. Pp- 49-66.
14. Richards A. J 1932: *Hunger and work in a savage tribe*. Routledge, London.

15. Schieffelein E 1977: *The sorrow of the lonely and the burning of the dancers* University of Queensland Press: St. Lucia.
16. Simoons F 1961: *Eat not this flesh*. University of Wisconsin: Madison.
17. Tongue R. K 1974: *The English of Singapore and Malaysia*. Eastern Universities Press: Singapore.
18. Titcomb M 1967: The foods of ancient Hawaii. *The conch shell Bishop Museum News*. Vol.4.
19. Young M. W 1971: *Fighting with food: leadership, values and social control in a Massim society*. Cambridge University Press: Cambridge.

Floral Diversity and their Conservation (2013), Editors: D.R. Khanna et al.
Published by Biotech Books.
ISBN: 978-81-7622-286-0 Pages: 69-74

8

Conservation and Propagation Strategy for *Podophyllum hexandrum* Royle

Rajendra P. Kala *, Kataku Ram, Neeta Gera & A.K. Sharma****
*Department of Forestry, Uttaranchal College of Technology and Bio Medical Sciences, Dehradun
Aushdheeya Vanaspati Van Sanstha - Uttaranchal, Dehradun
** Non Wood Forest Products Division, Forest Research Institute, Dehradun

The Himalayas are a veritable emporium of over 3,000 species. Many of the high altitude MAP species *viz. Achillea millefolium, Aconitum heterophyllum, Aconitum ferox, Berginia ligulata, Picrorrhiza kurrooa, Nardostachys jatamansi, Podophyllum hexandrum* have been ruthlessly extracted in Uttarakhand and other Himalayan states, without any concern for their regeneration by the scrupulous forces of the market.

There is an urgent need to develop cultivation protocols for rare, endangered and important medicinal plants so that techniques and economic packages can be transferred to the cultivators. A number of efforts have been reported in this regard.

Podophyllum hexandrum Royle. is one of the most important high altitude medicinal plant which is used as a remedy against different types of cancers and has an International market. The species has been banned from wild collection in India and has also been listed by the CITES as endangered.

The rhizome of this plant is the desirable part and is known to have comparatively longer period for maturity for therapeutic use and economic returns.

A simple and cost effective seed production / protection and nursery seed germination practice has been developed for *Podophyllum hexandrum.* The seed germination of 65-70 % has been achieved using different seed pre-treatments.

In nature, the mature seeds of this plant are rare to find because of the crimson colour of the fruit which attracts the birds and other wild animals leading to their unavailability for natural regeneration. The present study demonstrates a cost effective methodology to protect the mature fruits in nursery and germplasm collections.

Key words: *Ex-situ* Conservation; *Podophyllum hexandrum;* Seed Germination.

INTRODUCTION

Conservation of *Podophyllum hexandrum* Royle, an important medicinal plant in Garhwal Himalayas, Uttarakhand, India

- Himalayas encompass a great diversity of medicinal and aromatic plants, which possess many therapeutic values capable of curing disease ranging from intermittent fevers to malignant organs. Of late, the identification of a few plant species helpful in the treatment of cancerous ailments has lead to the status survey in the wild. One such plant species of the temperate zone, Podophyllum hexandrum suffers from the problem of its in situ conservation and as well ex –situ conservation for obvious reasons of its long gestation period and removal of its edible crimson coloured fruit even before maturity by wild animals, birds and also human beings. According to the IUCN classification *Podophyllum hexandrum designated* the status of **critically rare.** *Podophyllum hexandrum* Royle is a glabrous, succulent; erect herb. The Plant flourishes well as an under growth in the Fir forest, rich in humus and decayed matter in Himalayas. The plant loves moist and shady localities situated between 2500-4000 m. Amsl Seed germination very poor under natural conditions. Seeds under natural conditions have been found to germinate after remaining dormant for one year or two years, but If sown without the pulp, fail to germinate. Preliminary treatments like soaking them in water, or treating with sulphuric acid has no effects. Seed production orchards are the need of the time to propagate. Need based protection from predators will help boosting the production of seeds and consequently the production of roots for medicinal use.

The Indian *Podophyllum hexandrum* is superior to its American counter part, *Fodophyllum peltatum* in terms of its higher Podophyllotoxin content (4%) in dried roots in comparison to only 0.25% 0f *Podophyllum peltatum*

Thc rationale of the present study is making available *Podophyllum hexandrum* Royle seeds for propagation through ex situ cultivation. By making available seeds from cultivation fields , assisting in the conservation of wild populations. Increasing the availability of the authentic plant material for medicinal uses.

MATERIALS AND METHODS

The germination and seed production studies were made in the Kauntalani nursery, Deoban Block of the Chakarta Forest Division, Uttarkhand. Seeds were collected from Mundali block of Deoghar Range of Chakrata Forest Division. Seeds were collected during the first and second week of September on the maturity of the berries. Seeds were extracted manually from the fruits and the pulp was segregated from the seeds after washing with water and dried under shade. The air dried seeds weighed app. 5000 seeds/ Kg. These seeds were given chilling treatment for 30 days (Mist chamber).Germinated in two different environments

– Mist Chamber and Open field.

Mist Chamber-The potting mix (mist chamber) sterilised sand was used. The sterilization of the sand was done by roasting it on high temperature and after cooling it was placed in the germination trays. The Mist chamber temperature was maintained at 30^0 C during the day and 20^0C during the nights, Relative humidity of 60-70%.Irrigation done by sprinkling water after every one to three days.

Open field- FYM, sand and soil (1:1:2) was used for the raised nursery beds for the sowing of seeds in November and the spacing was kept at 30 cm distance. The depth of seeds was maintained at 1.5 - 4cm. The raised beds were covered with the thatch material. Light irrigation was applied. The beds were irrigated every one to three days till the first snow fall. The thatch material was removed at the end of February and start of March depending on the melting of snow in the area.

RESULTS AND DISCUSSION

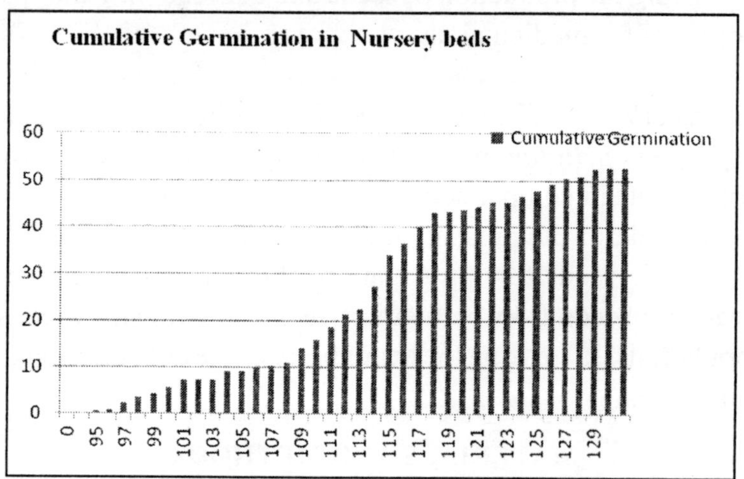

The germination started on the 94th day and continued up to 128th day.
The germination percent observed 53 %.
The germination completed in 128 days from the day of sowing of seeds.

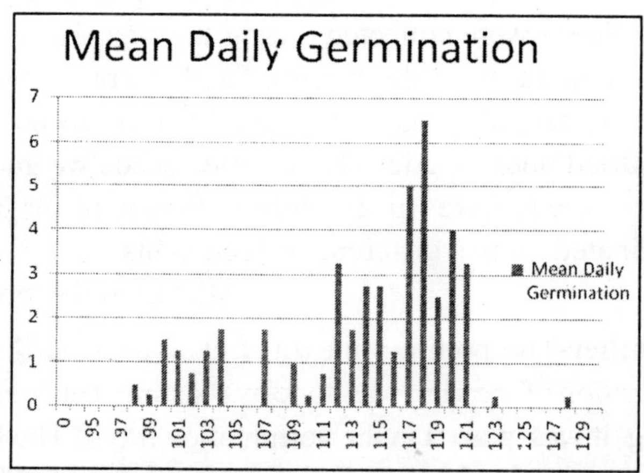

The germination percent observed is 53 % .
The maximum germination of 7 seeds was observed on 117th and 118th day.
Germination completed in 35 days.

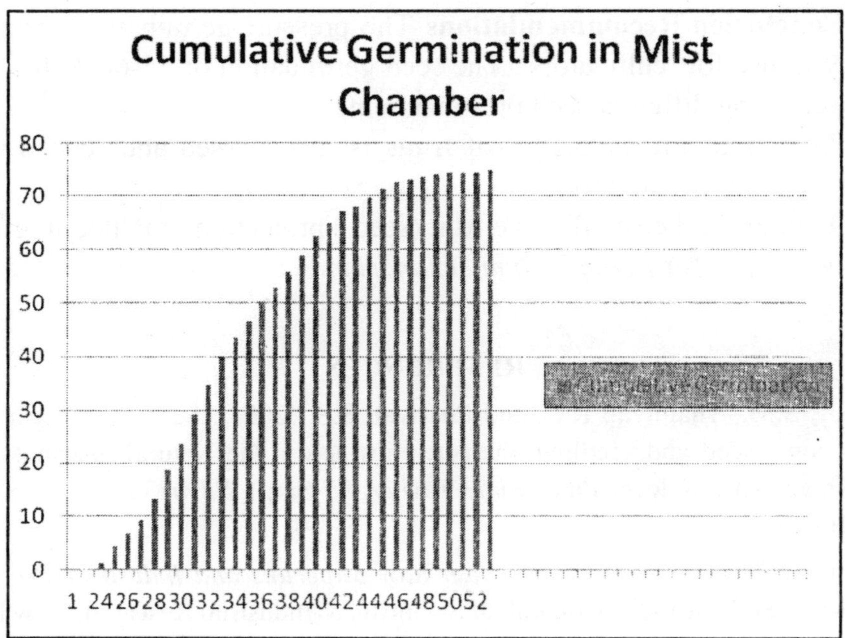

The germination started on the 23rd day and continued up to 48th day.
The germination percent observed 75%.
The germination completed in 48 days from the day of sowing of seeds.

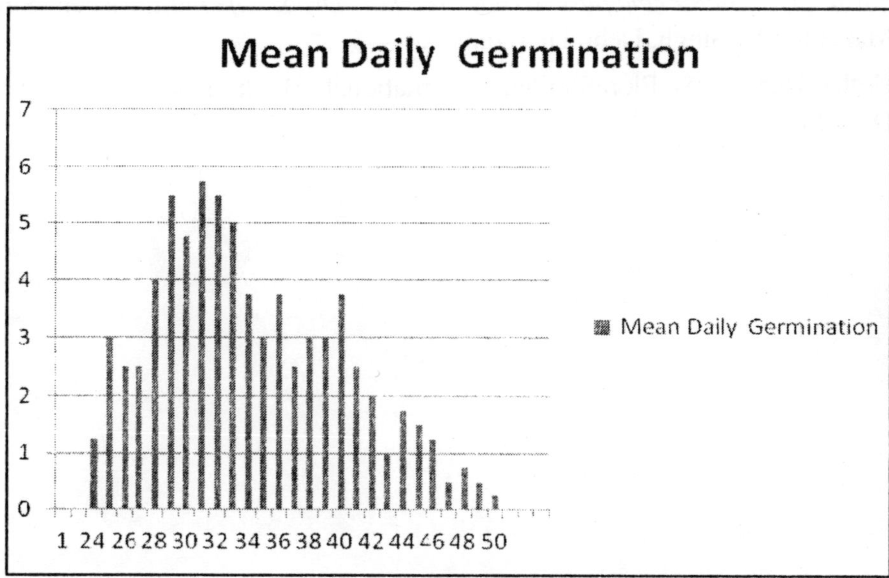

The germination percent observed is 75% .
The maximum germination of 7 seeds was observed on 28th ,30th and 31st day.
Germination completed in 26 days.

Conclusion/Recommendations The present germination technique is easy to use for cultivators . The seed germination of 55-60 % has been achieved using different seed pre-treatments.

Protection methodology of fruits is need based and cheap using locally available materials.

A simple and cost effective fruit (seed) protection methodology has been developed for *Podophyllum hexandrum.*

REFERENCES

1. Roy, S.M., Thapliyal, R.C. and Phartyal, S.S. (2004). Seed source variation in cone, seed and seedling characterstics across the Natural distribution of Himalayan low level Pine (*Pinus roxburghii* Sarg). *Silvae Genetica*, 53 (3): 116-128.
2. Anon.2005.*A Dictionary of Indian Raw Materials and Industrial Products*, The Wealth Of India. Council Of Scientific & Industrial Research, New Delhi .Vol.VIII:170-174.
3. Nadkarni, A.K.1976.Indian Materia Medica, Popular Prakashan, Bombay. Vol. I:904-905.
4. Sarin, Y.K.2008.*Principals Crude Herbal Drugs of India.*Bishen Singh MahendraPal Singh, DehraDun, India. :44-45.
5. Collett.Henry.1984.FloraSimilensis,Intrnational Book Distributors, Dehra Dun (India).:22

Floral Diversity and their Conservation (2013), Editors: D.R. Khanna et al.
Published by Biotech Books.
ISBN: 978-81-7622-286-0
Pages: 75-81

9

Effect of Conservation Trenches on Plantation Crop in Degraded Watershed in Kandhamal District of Orissa

Ch. Rajendra Subnudhi
Associate Professor, Department of Soil and Water Conservation Engineering, College of Agricultual Engineering and Technology, Orissa University of Agriculture Echnology, Bhubaneswar, Orissa, India

Kandhamal district situated in central part of Orissa receives an annual rainfall of 1396mm and this region is highly prone to soil and runoff loss due to heavy rainfall during kharif. A trial was conducted during 2001-04 to study the effect of conservation trenches on plantation crop. This trial was conducted on farmers field of Sudreju village of Kandhamal district under National Agricultural Technology Project(NATP, RRPS-7) with the following objectives.1.To conserve moisture for establishment of plantation crop. 2.To reduce erosion from upstream area.3 To increase production of timber, fruit species, fuel wood and fodder .The following treatments were tried.1.No treatment.2.Continuous V-ditches at 10m horizontal interval.3 Continuous V-ditches at 20m horizontal interval.4.V-ditches staggered at 5m horizontal interval. 5 V-ditches staggered at 10m horizontal interval.

Mango varieties Pusa Amrapalli was tried during kharif and during, rabi Black gram (PU-30) was tried in between mango rows. It is observed that in, cont. contour V-ditch at 10m interval rate of growth was 2.06 cm/month in case of Amrapalli , which is 46 %higher compared to control. The grain yield of niger, black gram & mustard are 33.4%, 23.5 % &26.6 % higher than control respectively. Though the cost of construction is little high it is recommended to practice contour V-ditch at 10m intervals, to conserve soil and moisture and to get more grain yield in degraded watershed of Kandhamal district.

INTRODUCTION

Although Kandhamal, receives around 1396 mm of rainfall, uneven distribution, heavy downpour of rain at times results in sudden high runoff which ultimately causes substantial soil loss. The uneven distribution of rainwater and movement of soil within the watershed, results in heavy loss to farmers. So conservation trenches for plantation crops helps to conserve the soil and moisture and ultimately improves grain yield of the farmers.

The objectives of the experiment are to conserve moisture for establishment of plantation crop ,to reduce soil erosion from upstream area and to increase production of timber, fruit species, fuel wood and fodder.

REVIEW OF LITERATURE

Samra, J.S.(2002) reported that renovation of terrace and plantation of fruit plants, timber plants improved biomass production, net returns, growth of crop, productivity, reduction of runoff in the range of 1.5-10.8 times, peak flow rate by 20 times& soil loss in the range of 1.2 to 5.2 times ,as well as water table rise. Subudhi *et.al* (1999) have reported that effect of vegetative barrier like Vetiver has increased the rice yield ,decreased soil loss and runoff compared to farmers practice. Arora *et.al* (2002) reported that there is a growing need for rain water management since 96 m ha out of 142 m ha of net cultivated land of the country is rainfed. Scientific use of these resources will definitely increase the productivity and conservation of resources like soil and water. Kumar (2002) reported that impact of different soil and water conservation techniques *viz. contour bunding*, terracing, land leveling, smoothening and gully plugging, sowing across the slope, vegetative barrier, increase the Kharif crops by 25-30 percent. .Establishment of vegetative barrier with mechanical measures were more effective in controlling soil erosion (3.8 t ha^{-1}) over conventional method

(9.64 t ha^{-1})and runoff thereby making more moisture available for crop growth. Anonymous (2003) reported that V-ditch at 10 m CCVD increased the crop yield significantly compared to no treatment.

MATERIALS AND METHODS

The study area lies in the Pila-Salki Watershed of Mahanadi Catchment. It falls under Sudreju revenue village of Khajuripada block in Phulbani district. As per Soil Conservation Department Govt. of Orissa, it is a part of watershed ORM 3-9-6-5. As per watershed map classification reported by the Orissa Remote Sensing Application Center (Department of Science & Technology, Govt. Of Orissa) the selected Micro-Watershed falls under Sub-Watershed No 17-07-31-01-01. This sub-watershed consists of parts of Survey of India Topographical Sheet Nos. 73D/2, 73D/6,73D/3 & 73D/7. However the Micro-Watershed under study falls only under Topo Sheet No. 73D/6. These Micro-Watersheds are located at a distance of about 10 km from Phulbani district headquarters on Phulbani-Sudrukumpa State Highway.

An on farm trial was conducted in the year 2001-04, at Sudreju under Dryland Agril Research Project, Orissa University of Agriculture & Technology, Phulbani, financed through National Agriculture Technology Project, Rainfed Rice Production System-7. Five following treatments were tested with 4 replication in randomized block design.

Treatments were;T_1-No treatment;T_2-Continuous V- ditches at 10m horizontal intervals.;T_3-Continuous V-ditches at 20m horizontal intervals. T_4-V-ditches staggered at 5 m horizontal interval.T_5-V- ditches staggered at 10m horizontal intervals.

The name of farmer is Kisore Pradhan. Mango variety Amrapalli was tried during Kharif in 5 meter spacing &Niger, Black gram and Mustard were tried during Rabi with 30cm spacing. Weather was favorable for all crops.

Disease & Pest: Mango hopper in all Mango varieties. Crop stand: Good.Slope: Field was contour surveyed and the slope was 4.15%.Soil loss was measured after the rainy season in the V-ditches, the soil was completely filled in 10m CCVD.So soil conserved was calculated as we know the size of the V-ditch before and after the rainy season.

RESULTS & DISCUSSION

Monthly rainfall is presented in Table-1.It is observed from above table that the year 2002 being a drought years, received only 74 % of rainfall, a deficit of 36 % from mean rainfall. But 2001 & 2003 are good years receiving 39.6% and 4% more than the mean annual rainfall respectively. The mean annual rainfall is 1396.14 mm. The fluctuation shows the rainfall is very erratic in all the three years.

Table 1: Monthly rainfall (mm) during 2001, 2002 &2003 and their deviation from mean.

Month	*Monthly normal*	*Actual in 2001*	*Deviation from normal, %*	*Actual in 2002*	*Deviation from normal, %*	*Actual in 2003*	*Deviation from normal, %*
January	9.18	-	-100	13.0	+41.6	0.0	-100
February	14.07	-	-100	-	-100	23.5	+67.0
March	21.70	56.0	+158.1	20.0	-7.8	12.5	-57.6
April	30.40	-	-100	32.0	+5.2	89.0	+192.7
May	57.48	48.0	-16.5	70.0	+21.8	7.0	-87.8
June	191.62	504.9	+163.5	149.0	-22.2	117.0	-38.9
July	353.62	797.6	+125.6	129.0	-63.5	237.0	-33.0
August	378.65	300.1	-20.7	329.0	-13.1	358.1	-5.4
September	218.57	124.7	-42.9	134.9	-38.3	350.1	+60.2
October	88.93	111.5	+25.4	11.0	-87.6	216.0	+142.9
November	27.48	6.9	-74.9	-	-100	0.0	-100
December	4.45	-	-100	-	-100	42.0	-843.8
Annual	1396.15	1949.7	+39.6	887.9	-36.4	1452.2	+4.0

Soil : The soil data has been presented in Table-2 it reveals that pH is low in top soil (5.42) compared to bottom soil (5.98)

Table 2: Soil Analysis Report

S. No.	*Name of the Farmer*	*Depth (cm)*	*Crop*	*pH (1:2.5)*	*EC (dsm^{-1})*	*OC (g/kg)*	*OM (g/kg)*
1	Kishore Pradhan	0-30	Mango	5.42	0.0174	5.62	9.67
2	Kishore Pradhan	30-60	Mango	5.98	0.042	3.26	5.61

Table 3: Yield ,plant height and moisture content and soil conserved in different treatments

Treatments	*Niger (q/ha) (2001-02)*	*Black gram (q/ha) (2002-03)*	*Mustard (q/ha) (2003-04)*	*Mean moisture Content (%) At 0-30 cm on weight basis during 2001-03*	*Mean rate of growth of mango (cm/month) during 2001-2003*	*Mean Soil conserved in ton/ha*
T_1- No treatment	2.33	6.12	4.17	3.67	1.22	0
T_2-Continuous V- ditches at 10m horizontal interval.	3.11	8.00	5.28	10.25	3.02	6.2
T_3-Continuous V-ditches at 20m horizontal interval.	2.44	7.12	4.85	5.59	2.47	3.2
T_4-V-ditches staggered at 5 m horizontal interval.	2.51	7.37	5.15	8.47	2.42	5.5
T_5-V- ditches staggered at 10m horizontal intervals.	2.49	7.25	5.00	7.02	2.50	3.1
SE (m)±	0.13	0.57	0.05			
CD (0.05)	0.39	NS	0.17			

Table-3 shows rate of growth of mango. The rate of growth is highest (3.02cm/month) in T_2-CCVD at 10 meter interval & lowest (1.22cm/month) in control from 2001-2003. The grain yield of Niger, black gram & mustard are 33.4%, 23.5 % &26.6 % higher than control respectively (Table.3). This may be due to more soil and water conserved at root zone of the crop as the moisture content in T_2 is more compared to all other treatments and lowest in control as there was no V-ditch (Table-3) The soil conserved in T_2 is 6.2 ton/ha followed by T_5 where soil conserved was 5.5 t/ha.Patil,P.P *et al.*(2004) has obtained similar result ,they got lowest soil loss (1.51 t/ha) and highest survival percentage of cashew nut plantation in Continuous contour trench compared to staggered trench (3.95 t/ha) and control (16.55 t/ha). So it can be cocluded that 10meter CCVD should be recommended for uplands of degraded watershed at Kandhamal district of Orissa.

SUMMARY & CONCLUSION

The present study reveals that grain yield of niger, black gram & mustard are 33.4%, 23.5 % &26.6 % higher than control respectively. Though the cost of construction is little high it is recommended to practice contour V-ditch at 10m intervals, to conserve soil and moisture and to get more grain yield in degraded watershed of Kandhamal district. It is observed that in cont. contour V-ditch at 10m interval rate of growth was 3.02 cm/month in case of Amrapalli, which is 46 %higher compared to control. Also we can conserve 6.2 t/ha of soil by 10m CCVD which is highest among all the treatments.

It can be cocluded that 10meter CCVD should be recommended for upland of degraded watershed of Kandhamal district of Orissa.

ACKNOWLEDGEMENTS

The authors acknowledge the help of Vice Chanceller, O.U.A.T., Director, CRIDA, Hyderabad & Dean of Research. O.U.A.T., Bhubaneswar for time-to-time guidance &financial help to carry out this project. The authors also acknowledge the help of D.L.A.P. staff of Phulbani & staff of CAET, OUAT, Bhubaneswar, who are helping for the success of the project.

REFERENCES

1. Annonymous (2003) Final progress report of NATP, RRPS-7, DLAP, OUAT, Phulbani.
2. Arora Dinesh & Gupta A.K.(2002) Effect of water conservation measures in a pasture on the productivity of Buffel grass. Proceedings of Indian Association of Soil & Water Conservationists, Dehradun conference held in 2001. pp-65-66.
3. Eswaran V.B.(2002)Wasteland development. Proceedings of Indian Association of Soil & Water Conservationists,Dehradun conference held in 2001.pp 17-19

4. Kumar munish (2002).Impact of soil & water conservation on erosion loss and yield of Kharif crops under ravenous watershed. Proceedins of Indian Association of Soil & Water Conservationists,Dehradun conference held in 2001.pp 301-303

5. Patil,P.P.,Gutal G.B.,Ganvir,B.N. and Bodake,P.S.,(2004)Soil and moisture conservation practices for the hill slopes in Western Ghat of Maharashtra. Extended abstracts of National Conferences on Resource Conserving Technologies for Social Upliftment.pp 122-124.

6. Samra,J.S.(2002)Watershed management a tool for sustainable production. Proceedings of Indian Association of Soil & Water Conservationists,Dehradu n conference held in 2001.pp 1-10

7. Subudhi C.R.,Pradhan,P.C.&Senapati,P,C.(1999)Effect of grass bund on erosion loss and yield of rainfed rice, Orissa, India, T.Vetiver Network.19:32-33.

Floral Diversity and their Conservation (2013), Editors: D.R. Khanna et al.
Published by Biotech Books.
ISBN: 978-81-7622-286-0 Pages:83-97

10

Biodiversity, Traditional Practices and Sustainability Issues of East Kolkata Wetlands

SAYAN BHATTACHARYA*, ANTARA GANGULI, SREEDIPA BOSE, SAUMIK PANJA, PALLABI BHOWMIK AND ANIRUDDHA MUKHOPADHYAY
Department of Environmental Science, University of Calcutta.
51/2, Hazra Road, Kolkata, India

The East Kolkata Wetlands, situated on the eastern fringes of the Kolkata city, India, form a part of the extensive inter-distributory water networks of the Gangetic Delta. It has been designated as 'wetlands of international importance' in November 2002 by the Ramsar Convention on Wetlands for its environmental benefits. The 12,741-hectare area is a part of the lower deltaic plain of the *Bhagirathi–Ganga* river system and is generally flat in nature. It serves as a 'Natural Kidney' for Kolkata, receiving 250 million gallons of anthropogenic wastewater daily and has the capacity to remove around 237 kg. of BOD per day. The resource recovery systems developed by the local people through the ages has saved Kolkata city from the costs of constructing and maintaining wastewater treatment plants. It is the largest ensemble of sewage fed fish ponds in the world and also very rich in plant and wildlife diversity and genetic resources. Livelihoods of the local inhabitants are distinctly linked to the wetland resources, and 74% of the working population depends on fish farming, agriculture and horticulture for their survival. The wetland is absolutely indispensible for the security of food and health in Kolkata. However, Kolkata's intensive urban expansion, industrial pollutants (by raw sewage, plastics and tannery effluents), siltation, weed infestation and the changed land use patterns are continuously damaging the ecological health of East Kolkata wetlands. Implementation of sustainable policies along with the awareness of the people can maintain the ecological equilibrium of these valuable water bodies, which, in turn, can enhance social, environmental and economic security of Kolkata in future.

Key words: Wetland, Biodiversity, Wastewater, Pollution, Sustainability.

INTRODUCTION:

Wetlands are ecotones or transitional zones that occupy an intermediate position between dry land and the open water bodies. Water is the main dominating factor for controlling the unique characteristics of wetlands. The three main features of wetlands are: (1) groundwater at the surface or within the soil root zones (2) specialized wetland vegetation (hydrophytes), and (3) hydric soils. They execute both the characteristics of the terrestrial and aquatic ecosystems and also have some unique characteristics of their own. In most parts, they are covered with water either temporarily or permanently. With the help of the incoming solar radiation and existing nutrients, varieties of plants can flourish in the wetlands, which, in turn, can support the lives of secondary and tertiary producers (Ghosh, 2005). India by virtue of its extensive geographical and climatological variations can support a rich diversity of inland and coastal wetlands which harbour rich biodiversity.

Ramsar is a city in Iran where the first World Convention on Wetlands was held on 2 February 1971. India has been a participating country in Ramsar Convention since 1981 and has identified important activities for the conservation and wise use of wetlands. The main focus of the first Ramsar Conference was largely on the role of wetlands as waterfowl habitat. Over last 30 years, the perception of Ramsar convention has undergone significant changes (Ghosh, 2005). The convention gradually followed the holistic outlook of wetland management that includes fisheries, tourism, bio-diversity conservation, waste recycling, pollution mitigation etc. The universally accepted definition of wetlands according to the Ramsar Convention is: 'Wetlands are area of marsh, fen, peatland or water, whether natural or artificial, permanent or temporary, with water that is static or flowing, fresh, brackish or salt, including areas of marine water, the depth of which at low tide does not exceed six metres' (Ramsar Convention on Wetlands, 2011).

The social, economic and ecological values of the wetlands are increasingly receiving due attention as their ecological contributions are significant for maintaining a healthy environment in multidimensional ways. They can perform numerous valuable functions such as nutrient recycling, water purification, ground water recharge and sustainable sources of drinking water, flood management, stabilization of coastal areas and soil, biogeochemical cycles functioning, carbon sequestration and also the sources of biological and genetic resources (Prasad *et al.* 2002). They retain water during dry periods and can keep the water table high and relatively stable. They can mitigate flood by trapping suspended solids and attached nutrients. Chemical conditions in the soils and water allow proper cycling of nutrients. The nutrients can be used by plants for growth or can be released into the atmosphere as well. Wetlands are also important feeding and breeding grounds for wildlife and can provide ecologically sound habitats for maintaining species diversity (Prasad *et al.* 2002). The negative interactions of man with wetlands during the last few decades have been of serious concern largely due to the rapid population expansion, industrial development, unsustainable resource exploitation and pollution. More than 50% wetlands of the world have been destroyed in the 20th century, especially in developing countries, largely due to industrialization.

The East Kolkata Wetlands, situated on the eastern fringes of the Kolkata city, India, form a part of the extensive inter-distributory water networks of the Gangetic Delta. It has been designated as 'wetlands of international importance' in November 2002 by the Ramsar Convention on Wetlands on the basis of wise use to produce a range of goods and services, sustainable and eco-friendly system of sewage treatment and a habitat for diverse flora and fauna including waterfowls (Ghosh, 2005).

LOCATION, HISTORY AND CLIMATE OF EAST KOLKATA WETLANDS

Kolkata can be considered a ecologically subsidized city for its two special geographical features. One is the existence of the Hoogly River on the west and the other is the vast low lying area on the east towards which the city is inclined and plays the role of a sink. The wetland area exists approximately between latitudes 22°25' to 22°40' north and longitudes 88°20' to 88°35' east (Kundu *et al.* 2008). The wetlands system is the example of the largest and perhaps the oldest integrated resource recovery practice of the world which based on a combination of agriculture and aquaculture practices. Over 20,000 low income and economically marginalized families depend on the various biological products of the wetlands including fish and vegetables for livelihood (Ghosh, 2005).

The East Kolkata Wetlands are basically a part of the mature delta of the river Ganges. The process of land formation in this region has largely been influenced through centuries by the Ganges river system where upland and tidal depositions continue to create land. The tributaries and distributaries of the river were active in this area many years ago. Before the establishment of Calcutta, the salt-water lake formed here was a backwater swamp and spill area of the *Bidyadhari* river. Before 1830, these low-lying regions were utilized for cultivation of brackish water fishes such as Bhetki (*Lates calcarifer*), *Parse* (*Mugil parsia*), *Bhangar (Mugil tade*) and Prawns *(Macrobrachum rosenbergii*) (Ghosh and Sen, 1987). Until 1830, the *Bidyadhari* river was an active navigation route which helped to connect Bay of Bengal with Kolkata (DEC, 1945). The tidal flow from the Bay of Bengal was the main source of salinity in water in these areas. By the end of the 19th century with the death of the *Bidyadhari* river, the process of natural silt deposition and raising the level of the spill area stopped completely. Anthropogenic activities further reduced the spill area and the channel beds. This led to the formation of the vast saltwater marsh to the east of Kolkata between Hugli on the west and the Kulti Gong on the east. During early days of settlements, the Britishes transportated goods for commercial purpose from the hinterlands of Bengal (like Assam, Chandpur, Khulna, Dhaka or Barisal). Over a period of time, to support the waste management systems of the Kolkata city, the East Kolkata Wetlands transformed from brackish water lakes to sewage fed fish farms and the large areas were converted for human settlements and agriculture development. With the diversion of city sewage and storm water into the salt lakes and the deterioration of river Bidyadhari there was a gradual decrease in salinity in the aquatic environment (Ghosh, 2005). For the expansion of the urban area in the post independence era, 1,000 ha of the northern portion of the wetlands were used for establishment of the Salt Lake City. The gradual reduction in area of the wetlands for economic and social purposes has reduced its capacity to recycle wastes and to attenuate floods. Despite having a large direct catchment of 1,625 sq km (including the basins of *Kulti, Piyali Bidyadhuree, Adiganaga* rivers), the wetland inflows are mainly controlled by the sewage generated from the Kolkata Municipal Corporation (KMC) (East Kolkata Wetland Management Authority, 2011).

The hot, monsoon climate of the East Calcutta Wetlands is largely governed by the North Himalayan mountains, the North-eastern Meghalayan plateau and the Bay of Bengal. The climate of the East Calcutta Wetlands broadly resembles that of Kolkata. Temperature remains high throughout the year in the wetlands with minor variations and the climate resembles the features of a tropical region with sufficient sunlight and vast water regime (Kundu, 2008).

BIODIVERSITY OF EAST KOLKATA WETLANDS

Wetlands are the rich sources of biodiversity. Varieties of plants and animals including birds, mammals, reptiles, amphibians, fish and invertebrate species can flourish in these ecologically favourable habitats. It has been estimated that freshwater wetlands hold more than 40% of the entire world's species and 12% of all animal species. Individual wetlands can be extremely important in supporting high numbers of endemic species.

Over last hundred years or more, the East Kolkata Wetlands have undergone drastic changes in floral and faunal diversity in response to changes in salinity regimes. The wetland once supported a rich floral and faunal diversity when connected to the freshwater and tidal regimes. Over the years, changes in salinity, land use-land cover, industrial pollution and other developmental activities, the species diversity has been reduced significantly (East Kolkata Wetland Management Authority, 2011).

Wetland is a complex and dynamic environment in which biological activity is mostly governed by microorganisms. The beneficial effects of wetland microorganisms can be executed in multidimensional ways such as nitrogen fixation, organic matter decomposition, breakdown of metabolic by-products and agrochemicals, enhancing the bioavailability of nutrients and essential metals. Studies revealed that East Kolkata wetlands have varieties of microbial population with diverse genetic characters including spore formation for stress adaptation, versatile enzymatic and metabolic activities, high pH and temperature tolerance, nitrogen fixing abilities and the capacity of bioremediation of different toxins, heavy metals and pollutants (Ghosh *et al.* 2007). A few examples of the microbial population include *Rhodococcus* sp., *Bacillus* sp., *Pseudomonas* sp., *Azotobactor* sp., *Aeromonas* sp. etc. (Ghosh *et al.* 2007). Plant roots grown in these areas can release compounds including simple sugars, amino acids, aliphatics, aromatics that can stimulate the growth and metabolism of specific microbial communities which accelerate the bioremediation processes.

In the early phase of the 20th century, 70 species of plants were recorded from various zones which were distributed within embankments (*Fimbristylus ferruginea, Suarda maritime, Acanthus illicifolius, Excoecaria agalocha, Avicennia officinalis*), main wetland (different species of algae, *Phragmites karka, Aegiceras majus, Typha elephantine* etc.) and terrestrial area (oligihaline and mesohaline shrubby plant species and several halophytic trees like *Sonnerata apetala, Avicennia officianalis*) (Biswas, 1927). Dasgupta (1973) recorded 97 plant species among which, 34 species were confined to saline water habitat. He also mentioned the existence of mangrove plants in these regions. A recent study, however, reported 106 aquatic plants belonging to 70 genera and 36 families (Ghosh and Ghosh, 2003). Among the vegetations, there are very few submerged vegetations remain in the core fishing area except the plankton communities.

East Kolkata wetlands harbour variety of economically important plant species. Some of these plants have rich medicinal values and have been used traditionally by the local communities over a long time. Besides several plant species such as *Bacopa monnieri, Enhydra fluctuans, Ipomea aquatica, Marsilea minuta* are used as vegetables by the local communities. *Cyperous rotundous, Phragmities karka* and *Typha angustifolia* are used by the local communities for thatching as well as for pulp, fiber and other uses. Several aquatic plants are used as green compost and manure apart from their utilization as fish food and for water purification (East Kolkata Wetland Management Authority, 2011).

Previously the fish fauna in the East Kolkata Wetlands system included both brackish water and fresh water forms. The low-lying regions with saltwater lakes were ideal for farming of brackish water fishes. Among the cultured fishes, *Bhetki* (*Lates calcarifer*), *Parse* (*Mugil parsia*), *Bhangar* (*Mugil tada*) and Prawns (*Macrobrachium rosenbergii*), etc. were common (Ghosh and Sen, 1987). Exotics *Clarius guriepinnus* and *Pangasius sutchi* which are banned species have also been recorded from the aqua culture farms. Indigenous species *Nandous nandus* and *Xenentodon cancila* were abundant during 1980s in these wetlands, but now no records found in support of their existence. Presently, in the *Bheris*, Poly-culture is practiced by the farmers and includes three species of Indian major carps (*Labeo rohita, Catla catla and Cirrhinus mrigala*) along with one minor carp (*Labeo bata*), three exotic carp (grass carp, silver carp and common carp) and two types of Tilapias. However, the presence of invasive exotic fish species is posing great threats to the native fish diversity (Ghosh, 2005).

The amphibian diversity in the wetlands is rather low and only 4 species recorded so far. The reptiles available include 19 species and are mainly water snakes, monitor lizard, common lizard and fresh water tortoise. The invertebrates, lower vertebrates including fishes and amphibians are the sources of food for the reptilian species.

Zoological Survey of India studied a total of 248 bird species (90 aquatic, 11 semi aquatic and 147 terrestrial species) in East Kolkata Wetlands. About 50% of the aquatic birds were reported to be migratory. However, the changed environmental conditions over the years have resulted in absence of larger species of birds like the Openbill Stork, Spoonbill, some species of ducks (e.g. Brahminy Duck, the Comb Duck), Red crested Pochard, Tufted Pochard etc. (East Kolkata Wetland Management Authority, 2011). Additionally, the predator birds like Brahminy Kite, Osprey and Laggar Falcon, Vultures etc. which were the common inhabitants in recent past are now no longer seen. The populations of Gadwals, Garganeys, Snipes, gulls, Terns, Egrets and Cormorants are also declining in population because of environmental and anthropogenic changes. It is assessed that 109 species of birds have become locally extinct, majority of which includes aquatic birds.

The mammal communities include marsh mongoose, an endangered one, which, borrows along the slopes of water bodies and eat fishes and aquatic snails in the wetlands. The species however, faces main threat due to habitat loss. Conversion of large parts of wetland areas into aquaculture ponds which are mostly devoid of vegetation is a major stress to foraging of this species. Besides, Common jungle cats, foxes, house rats, and mice are common inhabitants in these areas. Among the rare mammals, small Indian mongoose, palm civet, and small Indian civet are significant in and around the East Kolkata Wetlands.

WASTEWATER UTILIZATION AND WASTE RECYCLING IN EAST KOLKATA WETLANDS

Among the Ramsar sites of international significance, East Kolkata wetlands are unique for the wise use of sewage water, mainly for fish cultivation and irrigation for garbage farming. The wetlands consist of about 170 sewage fed fisheries that covers about 7000 acres of areas. The local name of the fish ponds is *Bheri*, which are shallow flat bottomed wastewater fed lagoon type of ponds and the size vary between 40–50 ha with 50 – 150 cm in depth (Roychowdhury *et al.* 2008). The entry of the increasing volume of sewages from Kolkata city into this area helps to decrease the salinity which is ideal for pisciculture. The Kolkata Municipal Corporation area generates about 600 million litres of sewage and waste water every day and more than 2500 metric tons of garbage (Roychowdhury *et al.* 2008). The wastewater is led by underground sewerage systems to the pumping stations in the eastern region and then pumped into open channels. The sewage and waste water is drawn into the fisheries of east Kolkata wetlands, where within few days, biodegradation of the organic compounds of the sewage and waste water take place. Retention of wastewater in the ponds before the initial stocking of fish for a considerable period of time allows bacterial population to metabolize the organic matters of the sewage and to decompose the organic wastes. The growth of these beneficial bacteria is supported by the algae that thrive in these shallow ponds in presence of sufficient sunlight. The biochemical reactions occurring in the wastewater ponds include photosynthesis, aerobic oxidation, organic acid formation, methanogenesis, nitrification and denitrification.

The basins of the *Bheris* are shallow which allows full vertical circulation of water to the surface where algal blooms occur. This is highly favorable for photosynthesis, as this depth gives a better ratio between pond volume and pond surface than a deeper pond. This helps to create sufficient oxygenation, which, in turn, can reduce BOD (Biological Oxygen Demand)

and faecal coliform load (East Kolkata Wetland Management Authority, 2011). The photosynthetic activity in the pond is one of the reasons for the purification of sewage. The water from the Raw Sewage Canal is drained into the *Bheris*, where it undergoes purification before being used for irrigation. The water hyacinths grown in these areas play major role in water purification and stabilization. They can leach out heavy metal ions from the contaminated water as their roots are efficient absorbers of heavy metals present in sewage water.

This system supports the recycling of the wastewater as well as the fish cultivation. The fishes maintain proper balance of the plankton population in the ponds and also convert the available nutrients of the wastewater into readily consumable form (fish biomass) for human consumption. Kolkata city receives about one third of its daily requirements of fishes from the sewage fed fisheries of the East Kolkata Wetlands (East Kolkata Wetland Management Authority, 2011).

Another system of waste recycling in the East Kolkata Wetlands is the use of garbage and waste water as manure and water for irrigation in paddy and for vegetable farming. The old solid waste dumping grounds have been converted into cultivable lands. The garbage farms can produce 150 tonnes of vegetables daily and the paddy fields can produce 16,000 tonnes of winter paddy varieties cultivated during post monsoon period annually. Different varieties of green leafy vegetables like *Amaranthus caudatus, Amaranthus blithum* and *Spinacia oleracea* are cultivated in these areas (Ghosh, 2005).

Livelihoods of the East Kolkata Wetlands communities are distinctly linked to wetland resources. About 74% of the local working population depends on fish farming, agriculture and horticulture in these areas. These creative and sustainable practices are developed by the local people and have been carried over generation after generation. These wetlands have been preserved and nurtured by the local communities, mostly belonging to the poorer and weaker sections of the society. These wetlands execute one of the biggest examples of low cost alternatives in municipal waste management that can ensure sustainable recovery of nutrients available in waste.

THREATS TO EAST KOLKATA WETLANDS: FEW OBSERVATIONS

In spite of being the only Ramsar site in West Bengal, the East Kolkata Wetlands are now facing several natural and anthropogenic stresses which altogether can lead to social, biological and economic degradations of the water bodies and adjoining areas. Some possible threats are discussed below:

A) Unsustainable use of groundwater in the East Kolkata Wetlands due to rapid urbanization, agricultural and industrial development has posed a risk of land subsidence. The subsurface geology of the area consists of quaternary sediments comprising a succession of clay, silty clay and sand layers of various grades. Groundwater occurs mostly under confined conditions except in those places where the top aquitard has been obscured due to the scouring action of past channels. Currently, the hydraulic head shows a falling trend and it may be accelerated due to further overuse of groundwater, which, in turn, may lead to land subsidence. The groundwater of the East Kolkata Wetland areas should be developed cautiously based on the groundwater potential to minimize the threat of land subsidence (Sahu and Sikdar, 2011) .

B) There has been a progressive shift in the land use within the wetlands leading to a gradual dominance of agriculture, which accounts for around 40% of the wetland areas. The areas under fish farms has reduced from 7,300 ha in 1945 to 5,842 ha in 2003. Construction of fish farms bunds and roads within the fish farms have further reduced the effective area under waterbodies to 2,481 ha. The gradual reduction in hydrological regimes within the wetlands has reduced its capacity to recycle wastes and attenuate floods.

C) There has been a rapid change in biodiversity associated with the wetlands due to changes in hydrological regimes and land use. Similarly, there has been significant loss of floral diversity, particularly those of mangroves and other brackish-water species. The local population depends on the vegetations of the wetlands for food, some of which are threatened due to environmental and anthropogenic stresses. The disappearance of these plant species will surely affect the food security of the low income local inhabitants in future.

D) Increasing concentrations of heavy metals in the sewage used in the wetlands attributed to unregulated discharge of industrial pollution poses a major threat both to the ecosystem as well as to the local communities. Several researchers have showed that some plants and animal species including fishes can bioaccumulate heavy metals like chromium, lead, mercury etc. from the wastewater and industrial effluents of the East Kolkata Wetlands, which can create health hazards (Chatterjee *et al.* 2006). However, some researchers also reported that the accumulation level is not significant enough so as create health hazards (Chattopadhyay *et al.* 2002).

E) Increasing siltation in the canals and fishponds has subsequently reduced the quantity of sewage flowing to the fisheries and made many of the fish ponds much shallower; which, in turn, has declined the fish production.

F) While environmentalists are making efforts for the preservation of the East Kolkata Wetlands, profit intensive people are increasing pressure for the right to develop areas for residential and industrial purposes. The wetlands are very close to Kolkata city, and The Eastern Metropolitan Bypass has made the area accessible and Salt Lake City apparently provides additional social and economic infrastructure. It is becoming difficult to protect the wetlands from developers and real estate agents. Public agencies have also shown a tendency to encroach upon the wetland areas for various developmental activities such as establishment of industries, commercial hubs or public utilities. It is increasingly becoming apparent that the existing legal provisions and agencies responsible for implementing them are unable to prevent such encroachment. Constructions in the wetland areas can eventually hamper the recharging of the groundwater and can also increase the incidence of flood. Development can also cause fragmentation of east Kolkata wetlands, which will increase social, ecological and environmental imbalances in local areas.

G) The sewage fed fisheries of the wetlands has been constrained due to inadequate management of water regimes, technology integration and weak marketing, post marketing and value addition opportunities. Management inefficiencies such as failure to properly maintain sluice gates and to run the pumping system, regulating the storm weather flow and the dry weather flow channels of the Kolkata drainage system in line with the requirements of farmers in these areas is gradually leading to collapse of the system. The current farm management systems are executing a biased outlook towards the large private farmers and against the small and medium size cooperatives.

SUSTAINABLE MANAGEMENT OF THE EAST KOLKATA WETLANDS: KEY ISSUES

The need of the hour is to stop further deterioration of the East Kolkata Wetlands and to protect and restore their original characters. Inappropriate understanding of the significance of these wetland practices has led to gradual loss of system efficiency. Based on assessment of hydrological, ecological and socio economic aspects and analysis of management practices, following key issues have been identified for the conservation and management of East Kolkata Wetlands:

A) For proper maintenance and functioning of the wastewater recycling process in the wetlands, all drainage channels and distributaries within the wetlands should be brought under a comprehensive action plan and restoration of the vital drainage structures is essential. Desiltation of the inflowing sewers can enhance water circulation and flushing in the wetlands systems. Improved waste segregation, proper handling of wastes and sustainable waste management strategies can help to mitigate pollution in the landfill areas.

B) The industrial effluents, especially those released from the nearby tanneries, can cause undesirable impact on the fish and vegetables grown in the wetlands. It is essential to identify such industrial units and to bring them under the strict regulations of pollution control. Common *in-situ* effluent treatment plants for the polluting industries can be effective for mitigating the pollution problem.

C) Farmers should be provided with a list of safe species that can be grown in the garbage farms in these areas and the extent of food chain contamination of heavy metals and industrial pollutants through different species should be evaluated in details for human health security. Focus should be given on proper identification and scientific uses of the bio-indicator species of these wetlands and the applications of the species in pollution management. Hyperaccumulators of the pollutants should be cultivated in the contaminated zones to remove the same. Restriction should be imposed for using insecticides, pesticides and other harmful chemicals in culture ponds of threatened species.

D) A detailed biodiversity and genetic database of the wetlands should be made, which, in turn, can give the scope for using varieties of resistant and tolerant genes for research purposes. Isolation and culture of microbial strains will be beneficial for implementing bioremediation strategies. Comparative vulnerability studies of the wetland species will help to implement preferential conservational strategies based on the degrees of sensitivities. Applications of biotechnological and microbiological research methods along with the applications of remote sensing and GIS are essential for these studies.

E) Populations of all breeding waterbirds species should be increased through improvement of breeding habitats and conditions. Proper understanding of population dynamics, feeding habits, and specific requirements of key bird species is essential for their conservation. Focus should also be given on the conservation of native and endemic species, including the fishes cultured in the fisheries.

F) Enhancing the water use efficiency in irrigation systems, diversification of cropping patterns in the wetlands areas and the implementation of rainwater harvesting structures in agricultural fields can stabilize the groundwater system and also can enhance the water security of the local people.

G) The management plans should focus on ecosystem conservation, sustainable resource development and livelihood improvement. Institutional development, communication, education and public awareness are also the key management components which will promote sustainability. Ecotourism development and enhancement of the social and economic security of the local people is essential for proper execution of the holistic outlook of wetland conservation.

CONCLUDING REMARKS

Unfortunately, in spite of important progress made in recent decades, wetlands continue to be among the world's most threatened ecosystems. Development as expected has increased consumerism in a small section of the society, but it has degraded and depleted invaluable natural resources like wetlands. Wetlands functions, values and attributes can only be maintained if the ecological processes of wetlands are allowed to continue functioning. East Kolkata Wetlands system is a unique example of environmental protection and development management that is in harmony with nature and benefits are achieved at a much lower cost. Knowledge about development and environment is not enough, rather the upsurge of the common people and pushing the principles of environmental security and justice are extremely important to conserve this important Ramsar site of West Bengal.

REFERENCES

1. Biswas, K.P. 1927. *Flora of the Salt Lakes, Calcutta.* Journal of Department of Science, University of Calcutta, 8, 1-48.
2. Chatterjee, S., Chattopadhyay, B., Mukhopadhyay, S.K. 2006. *Trace metal distribution in the tissues of Cichlids (Oreochromis Niloticus and O. Mossambicus) collected from wastewater fed fishponds in east Kolkata wetlands, a ramsar site.* International Journal for Ichthyology and fisheries, 36, 119-125.
3. Chattopadhyay, B., Chatterjee, A., Mukhopadhyay, S.K. 2002. *Bioaccumulation of metals in the East Calcutta wetland ecosystem.* Aquatic Ecosystem Health and Management, 5, 191-203.
4. Dasgupta.R. 1973, *Contribution of Botany of a portion of Salt Lakes, West Bengal.* Bulletin of the Indian Museum, 1, 36-43.
5. DEC. 1945. *History of the Gangetic Delta, Appendix 1a.* Report of the committee to enquire into the drainage conditions of Calcutta and adjoining area, Drainage enquiry committee, Government of West Bengal, Kolkata, India.
6. East Kolkata Wetland Management Authority. 2011. *Conservation and management plans of East Kolkata Wetlands.* Report to be submitted to the MOEF, Govt. of India.
7. Ghosh, D., Sen, S. 1987. *Ecological history of Calcutta's wetland conservation.* Environmental Conservation, 14, 219-226.

8. Ghosh, D. 2005. *Ecology and traditional wetland practice.* Worldview, Kolkata, India.

9. Ghosh, A., Maity, B., Chakrabarti, K., Chattopadhyay, D.J. 2007. *Bacterial Diversity of East Calcutta Wet Land Area: Possible Identification of Potential Bacterial Population for Different Biotechnological Uses.* Microbial Ecology, 54, 452-459.

10. Kundu, N., Pal, M., Saha, S. 2008. *East Kolkata Wetlands: A resource recovery system through productive activities.* Proceedings of Taal, 2007: The 12th World Lake Conference, 868-881.

11. Prasad, S.N., Ramachandra, T.V., Ahalya, N., Sengupta, T., Kumar, A., Tiwari, A.K, Vijayan, V.S., Vijayan, L. 2002. *Conservation of wetlands of India: A review.* Tropical Ecology, 43, 173-186.

12. Ramsar Convention on wetlands (the official website of the secretariat for the *Convention* on Wetlands, signed in *Ramsar*, Iran, in 1971). Accessed on October 7, 2011.

13. URL: http://www.ramsar.org

14. Raychaudhuri, S., Mishra, M., Salodkar, S., Sudarshan, M., Thakur, A.R. 2008. Traditional Aquaculture Practice at East Calcutta Wetland: The Safety Assessment. American Journal of Environmental Sciences, 4, 173-177.

15. Sahu, P., Sikdar, P.K. 2011. *Threat of land subsidence in and around Kolkata City and East Kolkata Wetlands, West Bengal, India.* Journal of Earth System Science, 120, 435–446.

Floral Diversity and their Conservation (2013), Editors: D.R. Khanna et al.
Published by Biotech Books.
ISBN: 978-81-7622-286-0
Pages: 99-109

11

Biodiversity Utilization and its Management: Important Role Play by Ethic People of the Thar Desert, Rajasthan, India

Parihar G.R.
National Desert Ecology Fellow
Assistant Professor, Department of Zoology, New Campus,
Jai Narain Vyas University, Jodhpur, (Raj.)

The biodiversity of Thar Desert, Rajasthan is different from the rest of the other part of the country and the local inhabitant, ethic people, plays important role for its utilization and management. The way of life of the local people is very hardy; they survive with minimum water availability, tolerating the temperature of +50^0 C along-with extreme hot ways, sand storm and sand clouds except in winter season. The source of lively-hood have chiefly been conditioned by developing dependency on desert fauna and flora of which utilizing them as food (during the time of drought and famine), using of wood as kitchen fuel, other parts of tree for formation/construction of their huts and for utilization of medicines extracted from different plant/tree species as well as animal species basing on their inheritant knowledge. The village elders have identified the natural resources area and nomenclatured as Orans, is named after local God. In this area only seasonal grazing of livestock are allowed and the products of its sold and the revenue generated is being used for development of Temple. The hunting of any animals and pollarding of living trees are strictly prohibited. Any one who violates these oral and unwritten rules is socially punished and ostracized. Similarly, some areas has been marked as Gauchars, is remained under control of the village Panchayat (local village administrative body) wherein only grazing, collection of dry wood, pods, seeds and gum etc. are allowed without interfering/harming to fauna and flora. This act and restriction help for management and enrichment of biodiversity in the Thar Desert of Rajasthan. The local inhabitants have developed knowledge and theme about forecast system and perception related to rains and better agriculture crops basing on animal behavior and changes in morphological characters in plants/tree is remarkable technique and can never be gleaned from the meteorological science. Especially, the Meghwal and Chaudhary community fellows having good knowledge then other community fellows on forecast system and perception. The paper discusses in detail on utilization of biodiversity and its management by ethic people of this region and the important role play by them in the Thar Desert. The data on plants and animals products utilization for different purposes by inhabitants are depicted in Tables 1- 4.

Key words: Biodiversity, Utilization, Management, Ethic people, Forecast, Thar desert.

INTRODUCTION

The biodiversity is considered as health and wealth of any one of the region, village forest or the country. Thus, the cultural base knowledge is very important for understanding the biodiversity for enrichment, utilization, management with maintaining the terrestrial ecosystem and biological productivity of that particular area. The form and pattern of population inhabiting in the Thar Desert is quit distinct as compare to those in other parts of the country. Besides, a sizeable proportion of the maximum population here leads a migratory, semi-nomadic or nomadic life in past and few of them are till today also. This may be the one of the reason to adopt nomadic as a way of life to cope up with the unfriendly climatic condition.

Presently, many of community people have stable/permanent life in Dhani (small village wherein same community/caste people reside) or in different villages (mixed community people reside).

Parihar (1997) summarized in detail on biodiversity and stated that it is an inherent quality of the entire living organism. It is essential for attaining sustainable gains in productivity of region, their cultural and social and food habits as well as agricultural priorities area within the desert are variable with static and sparse rainfall. Such a low population, productivity, erosion in the age old system and depletion of ecological stresses over the centuries have also resulted in ingenuities in land and water management system, such as the *Khadin* (small dug well in desert) and *Nadi* (small man made water reservoir in low land area in desert region), by census opinion the permanent pasture (or the *Orans* and *Gaucher*) arrow, the villagers are (under oral rules) kept out of cultivation, especially as an insurance against droughts or famine. The concomitant decline in pasture land and fallow lands along with good diversity are also expected to take place (Prakash, 1996). From the biodiversity point of view the Indian desert or Thar Desert some resulted on biodiversity of Thar which has been maintained and replenished over the centuries by the people themselves (Kaul, 1993; Khoshoo, 1994 and Rogers and Panwar 1988). To understand the biodiversity of the Thar desert, Rajasthan, we have to understand this region in its totality.

METHODOLOGY AND CASE STUDY

I. Field survey, selection of villages for study, interviews held with ethic people and informations collected based on their survivality, income sources, food habit, live-stocks in relation to utilization of biodiversity and taking proper due care for its management.

II. Study division in village institution was taken up in detail on *i.e.*, Oran land, Gaucher land, Fallow lands, Plough fields, Sand dunes, Forest enclosures, Gravel land and Waste lands.

III. The people of the desert are having fairly good knowledge on flora and fauna existing in the form of oral tradition and desert ethos in their region as well as environmental issues. The understanding of biodiversity and its uses in day to day life and prior to the monsoon time about its favorable/unfavorable condition, such type of ethical knowledge is more helpful to predict the rainfall in relation to their different crop production and on-coming problems due to nature *viz.* disease, famine/ drought etc.

Thus, according to the environmental conditions and productions they make their social future plans. Following are the some important factors taking in to account by local people:

1. Drought/famine forecast perception: Local inhabitants are very much aware of the coming one problem on drought /famine in the Thar desert, Rajasthan basing on their long year experience with co-relating the behavior of different animals and plants/trees. They declare Kal/Akal (Drought/famine) and making arrangement accordingly.

The grand natural disaster of the Thar desert is called Kal or Akal. One Kal is equal to 8 years. They also face the Kal intermittently during the year or year by year or overlapping years or after some interval of the years related to particular deficiency in this region is divided into three types namely (i) Ann (food grains) Kal: Food scarcity, (ii) Jal (water) Kal: Water scarcity and (iii) Tri Kal: Food, Water and Fodder scarcity. The Proverbs in their local language (Marwari) about Kal or Famines are very much popular and expressing by the local people in this region:

"Pag Pungal, sir Merta, udaraj Bikaner
Bhoola chooka Jodhpur, thavo Jaisalmer"

(Pungal, Merta, Bikaner, Jodhpur and Jaisalmer are the famous towns of the Marwar region of Rajasthan. Thus, term in words related to Akal condition in different towns will be indifferent: *Pag Pungal*- feet are in Pungal (serious akal in Pungal); *sir Merta*- head in Merta (more serious in Merta); *udaraj Bikaner*- belli (abdomen) in Bikaner (at least some food will be available); *bhoola chooka Jodhpur*- little bit good in Jodhpur (little good food will be available) and *thavo Jaisalmer*- akal will halt permanent in Jaisalmer means no food any edible things will be available. In other words akal in Jaisalmer means sentence to death of human beings and livestock).

"Rato bole Kag, Din ko bole Siyal
Ya gam ka dhan mare, ya pade Akal"

(If crow bird (Kag) is crowing in the mid night and Siyal (Jackal-*Canis aureus aureus*) is vocalizing in the day hours; either there will be the heavy deficiency of the food grains or will have to face the Akal (Famine) in coming days during the year or this year is going to be a drought year).

2. Perception based Rainfall forecast: The local people are having the predictions about rain fall with co-relating the different animals in this region and found to be almost correct. Few of them are mentioned here under-

(1) If *Uromastic hardwicky*: start sealing its barrow the region will receive heavy rain within 24 hours.

(2) If "viper" *Echis carinatus*_crowl on to the upper canopy of trees it sure that rain will be fall within 12 hour in that area.

(3) If peacocks, *Pava cristatus*-vocalised about at 3 am rain fall will be early in the morning.

3. Biomass productivity in year: They are also co-relating with plants/trees for better crops. These are-

(1) The almost all *Calotropis procera* "Aak" produces seed vessels (pods) in high quantity in that region/area will lead to higher production of *Sesanum indicum* (Til) during the year.

(2) In the year when *Salvodora alcodes* fruit are in plenty on the trees is co-related for good yield of agriculture crops.

(3) *Lasiurus sindicus* (Sevan grass) is in flourish condition is an indication of good crop of Gwar (*Cyamopsis tetraginoloba*).

RESULTS AND DISCUSSION

The local inhabitant inspite of holding large agriculture but production is very low from the cultivated field or extremely unstable due to erratic and scanty rains. These are the main reasons that their dependency has been diverted towards for utilization of biodiversity. They are not only utilizing the fauna / flora and their products even they have fully understood how to maintain and save biodiversity. I furnished here under my few observations on biodiversity utilization and its management at my own experience had in the Thar Desert and on discussions/interviews had with local people:

Gum as tonic: Many species of flora of this region are producing the Gum *viz.*, Desi babul (*Acasia nilotica*), Kumbat (*Acasia senegal*), Khejari (*Prosopis cineraria*), Israeli babul (*Acasia tortolis*), Angreji babul (*Prosopis juliflora*) and *Commiphora mightii* etc. is having the medicinal and good market value. These trees are protected by them on priority because these act as a source of revenue generation to the local people. This gum is mixing with other herbal products and made sweet dish (Goond ke ladoo) and mostly use in winter season as a special diet to maintain the physical health and for development of immunity power in the body. Similarly, the same is being used by the ladies after infant delivery as an especial tonic diet to regenerate the body power and for quality milk for infant.

Vegetation as food: Due to scarcity of water in this region the green vegetable production is almost nothing and the people of this zone searched out the sources from desert floras products *i.e.*, pods, seeds, flowers, fruits and burs etc. to use as dry vegetable since long back and using the same as on. The same dry vegetables have been popularized in many parts of the country and it has taken a very good market shape. Apart from utilizing as food the people are selling the same which help to improve their socio-economic condition. The considering the highly economic importance of these trees the inhabitant not only protects these trees, they further multiplying in the Thar Desert which is a good sign of biodiversity. The different vegetation use by them as food is presented in table-1.

Table 1: The products of different vegetation used as food by peoples in the Thar desert, Rajasthan and sales in the market.

Sl. No.	*Botanical name of tree, shrub, herb and grass*	*Common Name*	*Nature*	*Parts used*
1.	*Acacia nilotica*	Deshi babul	Tree	Pods
2.	*Acacia senegal*	Kumat	Tree	Pods
3.	*Colligonum polygonoides*	Phog	shrub	Flower, fruits
4.	*Capparis deciduas*	Ker	Tree	Pods
5.	*Cenchrus ciliarus*	Bhurat	Annual grass	Seeds, burs
6.	*Cenchrus setigerus*	Dhaman	Annual grass	Burs
7.	*Citrullus ianatus*	Matira	Annual herb	Seeds
8.	*Ciltrullus colocynthis*	Tumba	Perennial herb	Seeds
9.	*Indigofera cordifolia*	Bekar	Annual herb	Seeds
10.	*Lasiurus sindicus*	Sevan	Perennial grass	Burs
11.	*Prosopic cineraria*	Khejari	Tree	Seeds
12.	*Urochloa panicoides*	Kuri-gass	Annual grass	Seeds
13.	*Zizypus nummularia*	Bordi	Tree & shrub	Seeds, fruits
14.	*Dactylcteninum sindicum*	Ghantia	Annual grass	Seeds
15.	*Salvodora aleodis*	Mithi jal	Trees	Seeds, fruits
16.	*Cucumis callosus*	Kachri	Perennial herb	Fruits

Medicinal use: This region is not only desert but it is so desert in availability of almost all the civic facilities and availability of the medicines was far from their imagination in the past. Hence, the people prepare many medicines from the extract of plant species/ product *i.e.*, seed, flower or parts are and from animal species basing on their own long old experiences and inheritant knowledge. Presently about fifteen plants/trees being used in different diseases is presented in table-2. Similarly, many animal species are also being used for medicinal purposes is shown in table-3.

Table 2: Socio-cultural values of some important plant species for medicines.

S. No.	*Name of plant*	*Usable part of plant*	*Medicinal use and cultural value*
1.	*Indigofera cordifolia*	Seeds	For cure of pneumonia disease
2.	*Calitropis procera*	Leaves	For body pain
3.	*Colligonum polygonoides*	Floral tops	Worship of lord Shiva
4.	*Capparis decidua*	Fruits, floral and root sand	Joint disc disease and digestive system
5.	*Urochloa panicoides*	Whole plant	Increase milk in cows
6.	*Barleria acanthoides*	Leaves, floral of plant	Kidney stone, teeth disease
7.	*Blepharis sindica*	Seeds of plant	Epidemic use and health increasing
8.	*Euphorbia granualta*	Plant leaves	For urology problems
9.	*Solvedora persica*	Fruits and plant leaves	Prevent snake poison and leaves for body pain
10.	*Aerva nilotica*	Root and gum	For many medicinal use
11.	*Prosipis cineraria*	Pods and leaves called 'loong'	Vegetable and goat fodder, and 'Bishnoi community worship it for Lord Jamboji'
12.	*Acacia senegal*	Fruit and gum	Best vegetable and medicine
13.	*Crotalaria burhia*	Floral	Use for increase of memory and intelligence of mind power
14.	*Cenchural ciliaris*	Seeds	For improvement of child health
15.	*Cenchural setigerus*	Seeds and flower	Weight of bull induced as 'Boss'
16.	*Azadirachata indica*	Green leaves	For skin diseases and disinfection of houses against mosquitoes and house fly

Table 3: Socio-cultural values of some important animal species for medicines.

S. No.	*Name of Animal*	*Usable part*	*Medicinal use and cultural value*
1.	*Chalcophora fortis*	Thorax	Ornamental purpose, toys, silver and gold ornaments
2.	*Uromastix hardwicki*	Muscles and fat	Cough, body pain and use for epidemic diseases
3.	*Varanus bengalensis*	Skin and muscles	Used as food for camels for gaining energy
4.	*Naji naja naja*	Whole animal	Animal is symbol of local Lord 'Gogadev'
5.	*Streptopelia decaoto*	Flight muscle	For treatment of night blindness
6.	*Vulpes vulpes pusilla*	Skin, flesh and bones	Caps, burnt bones are used for boil and injuries by peoples, camel treated for pneumonia
7.	*Canis aureus aureus*	Flesh and blood	Food and blood for tuberculosis treatment
8.	*Hemiechinus auritus collaris*	Fat	Infection of joint and lumbago
9.	*Hystrix indica indica*	Fat of wills	Rheumatic pains and other infection. Pen holder and boring instruments
10.	*Gazella bennetti*	Flesh and urine	Used for whooping cough
11.	*Lepus nigricollis*	Blood	Children's diseases
12.	*Antilope cervicapra rajputani*	Whole animal	Symbol used by Bishnoi community for Lord Jamboji
13.	*Francolinus pondicerianus*	Flight muscles	Food and pneumonia disease
14.	*Funambullus pennati*	Flesh	Cough for children's

Kitchen fuel wood: It is well known facts that in such a remote area no any other sources are available for kitchen fuel. Even their economic condition is also not allowed to arrange other alternative facilities. Therefore they are using dry wood of many different trees and plants as kitchen fuel by local inhabitant and this is the only source for them. The dry wood of many different trees and plants use for kitchen fuel is enlisted in table-4.

Hut/Thatch house: Most of the local people are constructing the huts and thatch houses in desert and rural area with using the dry twigs, grasses and straw (millet dry stalk) for roofing purposes because the other materials are not available. Moreover, they can not afford wherever it is available. These types of huts and thatch houses are not only in economic range but very good for climatic conditions which always remain cool in extreme hot and even materials is transferable. The different trees and plants use to construct huts/thatch house is depicted in table-4.

Table 4: Different trees/surbs/grasses used by local people as kitchen fire wood (KFW) and for hut/thatch house (HTH) in the Thar desert, Rajasthan.

S. No.	***Botanical name of tree and shurb***	***Local name***	***Nature***	***Kitchen Fire Wood (KFW) Hut/Thatch House (HTH)***	
1.	*Acacia nilotica*	Deshi babul	Tree	KFW	HTH
2.	*Acacia tortilis*	Israely babul	Tree	KFW	---
3.	*Aerva persical*	Bui	Shurb	KFW	---
4.	*Calligonium polygonoides*	Phog	Shurb	KFW	HTH
5.	*Calotropis procera*	Aak	Shurb	KFW	HTH
6.	*Capparis decidua*	Keg	Tree	KFW	HTH
7.	*Clerodendurm phlomidis*	Arna	Shurb	KFW	HTH
8.	*Cordia gharaf*	Goondi	Tree	KFW	---
9.	*Crotolqria burihia*	Sunniya	Shurb	KFW	HTH
10.	*Greuuia tanex*	Gangani	Shurb	KFW	---
11.	*Heliotropium rarifloram*	Kharsunniya	Shurb	KFW	---
12.	*Lycium barbarum*	Morali	Shurb	KFW	---
13.	*Mimosa hamata*	Jinjani	Shurb	KFW	---
14.	*Prosopis cineraria*	Khejari	Tree	KFW	HTH
15.	*Prosopis juliflora*	Angrezi babul	Tree	KFW	---
16.	*Salvodora oleoides*	Mithi jal	Tree	KFW	---
17.	*Zizyphus nummularia*	Bordi	Tree	KFW	HTH
18.	*Acacia senegal*	Kumbat	Tree	---	HTH
19.	*Leptidenia pyrotechnica*	Khimp	Shurb	---	HTH
20.	*Panicumturgiderm*	Murat	Grass	---	HTH
21.	*Lasiurus sindicus*	Sevan	Grass	---	HTH
22.	*Pennisetum typoiderm*	Bajri Doka	Food grain	---	HTH

Worship of trees: It is recorded that the Hindu community people worship the many trees *i.e.*, Peepal (*Ficus religiosa*), Neem (*Azadirachta indica*), Khejari (*Prosopis cineraria*), Mithi Jal (*Salvadora oleoides*), Ker (*Capparis decedua*) and many more other trees and shrubs species also worship that directly help to maintain the tree vigor and their further regeneration as well as ecosystem.

CONCLUSION

The present study clearly demonstrate that a fairly large number of endemic animals and plants act as insurance of survival. Peoples have also learnt to live with them. A variety of insects, reptiles, birds and mammals which support human survival are quite conspicuous which need to grow further in spite of some soil erosion in the system, and the uses of plants to predict weather or the presence of animals or their absence and their survival linkages are important for the welfare of human society. Now, it needs empirical approach on this subject.

Further, on these above said observations and personally I had the close experience in the Thar Desert of Rajasthan that how people of this region are surviving in such a hard and in unfriendly climatic conditions along with facing the many problems to meet out their minor and basic need for their survivality by utilizing of biodiversity with taking the proper due care for its management. This is the only example in the country and it will be injustice with them to match/compare with any one part of our country. These are the real people come blow poverty line (BPL). Hence, it is requested to Government to take utmost care to uplift them and to improve their socio-economic conditions on priority.

ACKNOWLEDGEMENTS

I am grateful to Dr. D. Mohan, Professor and Head, Department of Zoology, JNV University, Jodhpur for his keen interest and support in this work. I am also grateful to Dr. L.S. Rajpurohit, Professor, Department of Zoology, JNV University, Jodhpur for his willing support and his opinion on village institutions. My thanks are due to Mr. H.R. Bania, Scientist, Central Silk Board: Govt. of India (on study leave), Department of Zoology, JNV University, Jodhpur for sharing of his knowledge on desert life and suggestions. Thanks are also due to Shri Kailash, Research

Scholar, Department of Zoology, JNV University, Jodhpur for typing of this manuscript.

REFERENCES

1. Koul, R.N. 1993. Strategies for the biodiversity in Thar desert. In: *Indo- British workshop in Biodiversity* (IBWB), New Delhi, pp. 91-103.
2. Khoshoo, T.N. 1994, India's biodiversity: Tasks ahead. *Current Science*, 67: 577-587.
3. Rodgers, W.A. and Panwar, H.S. 1988. A report: Planning wildlife protected area network in India. *Wildlife institute of India*, Dehra Dun. Vol. 1.
4. Parihar, G.R. (1997), Thesis: "*Studies on the biodiversity of some selected area in the Desert*", Department of zoology, Jai Narain Vyas University, Jodhpur (Rajasthan), India, pp. 1-148.
5. Prakash, I. 1994. Biodiversity conservation in the Thar desert. *Indian Forester*, 120: 873-879.

Floral Diversity and their Conservation (2013), Editors: D.R. Khanna et al.
Published by Biotech Books.
ISBN: 978-81-7622-286-0 Pages: 111-116

12

Anticipated Performance Index (API) of some Tree Species Grown in Aurangabad City

CHAVAN B. LAND SONWANE N. S.
Department of Environmental Science
Dr.Babasaheb Ambedkar Marathwada University, Aurangabad

Vegetation affects local and regional air quality by altering the urban atmospheric environment. The trees grown in urban area affects the air quality in terms of reduction in air temperature, removal of air pollutants, energy effects on buildings and emission of volatile organic compounds. The plant response can be monitored by analyzing the plant extract for various parameters like chlorophylls, ascorbic acid, pH and water content. In the present investigation anticipated performance index (API) of four tree species is evaluated. *Azadirachata indica* and *Mangifera indica* were the tree species having good API values (4 of each)because of their biological and socio-economic uses while *Polyalthia longifolia* and *Dalbergia sissoo* were judged to be moderateand poortree species having 3 and 2 API values respectively.

Key words: API, APTI, Biological, Socio-economic, Tolerance.

INTRODUCTION

Air is one of the most important elements for all living beings. It is one of the important natural resource on which all life depends. Clean and pure air is needed for the healthy life of all living beings including plants. Nowadaystheatmosphere is being polluted;impure air affects human health as well as otherenvironmental resources such as water, soil, and forests. Ultimately air pollution affects adversely the process of development at local as well as global level. At the present condition crowded cities like India with increasing numbers of industries, intensive transport networksand high population density have became a major source of air pollution.

All over the world developed, developing and underdeveloped countries are experiencing rapid growth. This growth isoccurring at a considerable, and often increasing, economic and social cost,without implementing environmental policy and action. Increasing number of people, industry, and motor vehicles cause air pollution which poses aserious environmental threat in urban areas. The world health organization (WHO) and some other international agencies have identified urban air pollution as a serious concern.

Trees in city area are facing the adverse impacts of pollution. The leaf is the most sensitive part to air pollutants. Affected plants show some common effects such as decrease in chlorophyll content, Inhabitation in photosynthesis and decreasing plant growth (Davision and Blakemore, 1976). Air pollution tolerance index defines capability of plant to combat against the air pollution. Plants responses towards air can be assessed by air pollution tolerance indices (Sing and Rao, 1983). The present study was planned to access the tolerance capacity and anticipated performance index of common trees grown in and Aurangabad city.

MATERIALS AND METHODS

Method for biochemical parameters:

Variousbiochemical parameters like Chlorophyll content (Maclachlan and Zalic, 1963), ascorbic acid content (Keller and Schwager, 1977), leaf extract pH (Singh and Rao, 1983) and relative water content (Singh, 1997) were done from the collected leaf samples.

ir pollution tolerance Index (APTI)

t was calculated byfollowing the method (Singh and Rao, 1983).

$$APTI= \{A\,(T+P)+R\}\,/\,10$$

Where: A is ascorbic acid content (mg/g), T is total chlorophyll (mg/g), is pH of leaf extract and R is relative water content of leaf (%).

PI (Anticipated performance Index):

ombining the results of APTI values with some relevant biological and ocio-economic characters (Plant habitat, canopy structure, type of plant, aminar structure and economic value) the API can be calculated for different pecies. Based on these characters, different grades (+ or -) are allotted to lants. Different plants are scored according to their grades. The criteria sed for calculating the API of plant species is given in following table.

Grading Character allotted	Pattern of Assessment	Grade
a) Tolerance (APTI)	8.00 – 12.00	-
	12.01 – 16.00	+
	16.01 –20.00	++
	20.01 – 24.00	+++
b) Biological and Socio-Economic:		
1) Plant habitat	Small -	
	Medium +	
	Large ++	
2) Canopy Structure	Sparce/Irregular/ Globular-	
	Spreading crown/ Semi dense +	
	Spreading dense	++
3) Type of tree	Deciduous -	
	Evergreen +	
4) Laminar Structure – Size:	Small -	
	Medium +	
	Large ++	
Texture:	Smooth-	
	Coriaceous+	
Hardness:	Delineate	
	Hardy +	
5) Economic value-	Less than tree uses	
	Three to four uses+	
	Five or more uses	++

S. No	Score (%)	Assessment category
1	Up to 30	Not recommended
2	Up to 31-40	Very poor
3	Up to 41-50	Poor
4	Up to 51-60	Moderate
5	Up to 61-70	Good
6	Up to 71-80	Very good
7	Up to 81-90	Excellent
8	Up to 91-100	Best

RESULTS AND DISCUSSIONS

Anticipated performance index (API) of some plant species:

Table 1(a): Assessment parameters

S. No.	Local Name	Scientific Name	APTI	Tree Habitat	Canopy Structure	Tree type
1	Neem	*Azadirachta indica*	+++	++	+	-
2	Mango	*Mangifera indica*	++	++	+	+
3	Ashoka	*Polyalthia longifolia*	++	+	+	+
4	Sheesham	*Dalbergia sissoo*	++	++	+	-

Table 1(b): Assessment parameters

Size	Texture	Hardness	Economic Value	Total Plus	% Scoring	API Grade
+	+	+	++	11	69	4
+	+	+	++	11	69	4
+	+	+	+	09	56	3
-	-	+	+	07	44	2

Assessment of plant species:

S. No.	Local Name	Scientific Name	Total plus (+)	% Scored	API	Assessment
1	Neem	*Azadirachta indica*	11	69	4	Good
2	Mango	*Mangifera indica*	11	69	4	Good
3	Ashoka	*Polyalthia longifolia*	09	56	3	Moderate
4	Sheesham	*Dalbergia sissoo*	07	44	2	Poor

Plant species were evaluated for various biological and socio-economic as well as some biochemical characters. These parameters were subjected to a grading scale to determine the anticipated performance of plant species as described by (Sing and Rao, 1983). API was calculated using the assessment parameters with respect to grading characters using a summation of the anticipated performance of the plant species. *Azadirachata indica* and *Mangifera indica* was the tree species having good API value with respect to the biological and socio-economic uses while *Polyalthia longifolia* and *Dalbergia sissoo*were judged to be moderate and poor tree species respectively.

Azadirachataindica found to be most tolerant tree species among all other species. It has dense canopy. The economic and aesthetic value of this tree species is also well known.*Mangiferaindica* is also another tree species having good API and tolerant to air pollution. It is an evergreen tree having dense canopy, rough leaves and valuable for aesthetic and economic uses.*Polyalthia longifolia* is also evergreen tree but its economic uses are less so the API value calculated is moderate. *Polyalthia longifolia* is an evergreen tree but its economic uses are less so the API value calculated is moderate. *Dalbergia Sissoo* is a deciduous tree and the size is medium texture is smooth so the API value calculated is poor and not recommended for the planting nearby roads.

REFERENCES

1. Davision, A.W and Blakemore (1976): J. In: Effects of air pollutions on plant (Ed.T.A.Mansfield). *Cambridge University press, UK*: PP-209.
2. Singh, S.K. and Rao, D.N (1983): Evaluation of plants for their tolerance t air pollution.In: *Proceeding symopsium on Air pollution control*.Vol.1.India Association for Air pollution control,New Delhi, India. PP-218 224.
3. Machlachlan S and Zalic, Plastid (1963): structure, chlorophyll concentratio and amino acid composition of a chlorophyll mutent of barley.*Can.J.Bot.* 41 PP-1053-1062.
4. Keller, T and Schwager, H. (1977): Air pollution and Ascorbic acid.*Eur J.forPathol.* 7: PP-338-350.
5. Singh, A. (1977): Practical Plant Physiology. Kalyani Publishers. New Delhi.

Floral Diversity and their Conservation (2013), Editors: D.R. Khanna et al.
Published by Biotech Books.
ISBN: 978-81-7622-286-0 Pages: 117-123

13

Antibacterial Activities of Green Algae (*Chlorophyceae*) from Narmada River

Shailendra Sharma, Taniya Sengupta, Kapil Sunar, Garima Solanki, Mitashi Tanwar
PG Department of Biotechnology, Shri Umiya Girls College, Mandleshwar, (M.P.) India.

The objective of this paper is to isolate green algae "*Chlorophyceae*" from Narmada River and to study its antimicrobial activity. The isolated algae were of two different types one was the fresh algae and the other was the dried form collected from the shore line area of Narmada River. Ethanol, chloroform, water, glycerol, acetone, formaldehyde, phenol and acetic acid crude extract from algae were obtained and tested against one Gram's Positive bacteria "*Bacillus subtilis*" and another against Gram's Negative bacteria "*Escherichia coli*". The maximum antibacterial activity was recorded from formaldehyde extracts of both types of algae against both bacteria. No antibacterial activity was recorded from water extract. The minimum activity was recorded from dried algae extract against *Bacillus subtilis* as compared to fresh one whereas the dried extract was pointed out to give maximum inhibition against *Escherichia coli* as compared with the fresh one. The present finding indicates the need for detailed work on the aspect, especially with reference to antimicrobial activities against different pathogenic bacteria.

Key words: Chlorophyceae, *Escherichia coli, Bacillus subtilis*, Antibacterial activities, Narmada River.

INTRODUCTION

The Narmada is a river in central India and the fifth largest river in the Indian subcontinent. It forms the traditional boundary between North India and South India and flows westwards over a length of 1,312 km (815.2 mi) before draining through the Gulf of Cambey (Khambat) into the Arabian Sea, 30 km (18.6 mi) west of Bharuch city of Gujarat.[1] It is one of only three major rivers in peninsular India that runs from east to west (largest west flowing river) along with the Tapti River and the Mahi River. It flows through the states of Madhya Pradesh (1,077 km (669.2 mi)), Maharashtra, (74 km (46.0 mi)) – (35 km (21.7 mi)) border between Madhya Pradesh and Maharashtra and (39 km (24.2 mi) border between Madhya Pradesh and Gujarat and in Gujarat (161 km (100.0 mi)). The Narmada basin, hemmed between Vindya and Satpuda ranges, extends over an area of 98,796 km2 (38,145.3 sq mi) and lies between east longitudes 72 degrees 32' to 81 degrees 45' and north latitudes 21 degrees 20' to 23 degrees 45' lying on the northern extremity of the Deccan Plateau. The basin covers large areas in the states of Madhya Pradesh (86%), Gujarat (14%) and a comparatively smaller area (2%) in Maharashtra. In the river course of 1,312 km (815.2 mi) explained above, there are 41 triburaries, out of which 22 are from the Satpuda range and the rest on the right bank are from the Vindhya range.

Algae are large and diverse group of simple typically autotrophic organisms, ranging from unicellular to multi cellular forms. They have their pigments localised in membrane bounded intracellular bodies (Plastids) have no vascular system and do not develop from an embryo. Algae are an extremely important species they can produce more oxygen then all the plant in the world. They form an important food source for many animals. Thus they are at the bottom of the food chain with many living thing depending upon them. *Chlorophyceae* have *primary* chloroplasts, *i.e.* the chloroplasts are surrounded by *two membranes* and probably developed through a single endosymbiotic event. Green Algae have chloroplasts with chlorophyll *a* and *b*. Higher plants are pigmented similarly to Green Algae and probably developed from them.

To date, there is moderately an assortment of reports commerce with the antibacterial activity of solvent extract attained from freshwater algae. Antimicrobial activities in freshwater algae have been reported widly over for decades by various scientist (Debro *et al.*;1979, Barchi *et al.*; 1984, De Mule *et al.*; 1991, Ishida *et al.*; 1997, Ozdemir *et al.*;2001, Noaman *et al.*;2004). An effort has been derelict in the present study to explore the antibacterial activities of freshwater algae of Narmada River on both Grams positive and Grams negative bacteria.

MATERIALS AND METHODS

The green algae was collected from the shore line areas of Narmada River from Mandleshwar in two forms one is the fresh and the other war the dried form. The plants were washed in tap water and cleaned of all epiphytes and other attached material and salt as soon as possible after collection. Then the plants were air dried for few days in the shade.

2 grams of both fresh and dry algae were soaked into 10ml of each solvent (Acetic acid, acetone, chloroform, ethanol, formaldehyde, glycerol, phenol and water) for five days. The resultant crude extracts were filtered. The crude extracts were kept in 4° until testing their antibacterial activities against Gram's negative Bacteria *Escherichia coli* and Gram's negative Bacteria *Bacillus subtilis* were done. For testing the antibacterial activities Plates of Muller Hinton (Himedia) have been used.

On completion of 24 hrs incubation the zone of inhibition were measured in each plate for both the kind of green algae.

Table 1 shows the inhibition zone for each extract prepared from fresh green algae. Algae itself in water does not release any extracellular components for inhibiting the growth of bacteria and in organic solvents except acetone and glycerol the fresh green algae shows the activities against *Escherichia coli.* Against *Bacillus subtilis* the fresh algae also shows remarkable inhibition except chloroform and Glycerol as a solvent. Maximum inhibition had seen in both bacteria from formaldehyde extract that is for *Escherichia coli* and *Bacillus subtilis* 6.5 and 6.0 respectively (Fig 1).

RESULTS AND DISCUSSIONS

Table 1: Antibacterial activity of fresh green algae Chlorophyceae obtained from Narmada Shore line areas.

S. No	Extract prepared in	Zone of Inhibition (cm)	
		Escherichia coli	*Bacillus subtilis*
1	Acetone	00	0.8
2	Acetic Acid	2.8	3.0
3	Chloroform	2.0	00
4	Formaldehyde	6.5	6.0
5	Glycerol	00	00
6	Phenol	4.2	3.0
7	Water	00	00

Table 2: Antibacterial activity of Dried green algae Chlorophyceae obtained from Narmada Shore line areas.

S. No	Extract Prepared in	Zone of Inhibition (cm)	
		Escherichia coli	*Bacillus subtilis*
1.	Acetone	1.2	00
2.	Acetic Acid	2.3	2.6
3.	Chloroform	1.5	0.8
4.	Formaldehyde	7.2	7.8
5.	Glycerol	3.6	3.4
6.	Phenol	3.8	00
7.	Water	00	00

Table 2 shows the results for the dried form of green algae against both selected bacterial species. Alike fresh form the dried form of the green algae also did not show any anti bacterial activities. The other organic extracts show efficient inhibition against *Escherichia coli*, whereas in case of *Bacillus subtilis* the results were positive except acetone and phenol extracts. Here also the most effective results were given by formaldehyde extracts against *Escherichia coli* and *Bacillus subtilis* as 7.2 and 7.8 respectively (Fig 2).

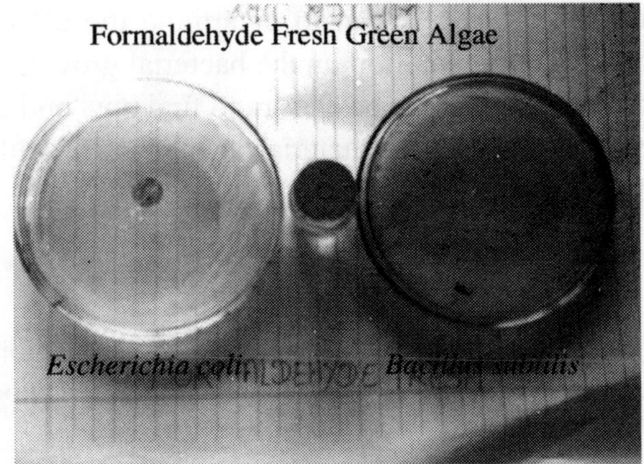

Fig. 1: Zone of inhibition against both types of bacteria by Formaldehyde fresh green algae extract.

Fig. 2: Zone of inhibition against both types of bacteria by Formaldehyde dried green algae extract.

On comparing it has been concluded that Formaldehyde extracts proved to be most acceptable antibacterial agent against pathogenic bacteria. Another thing which have been concluded was that the fresh algae extract was much effective on Gram's negative bacteria (*Escherichia coli*) and the dried form shows more response against Gram's positive bacteria (*Bacillus subtilis*).

Further work needs to be done to determine the effect of different solvents used for extract preparation on the bacterial growth and also other different types of bacteria have to be chosen to find how and to what extent these extracts can be used as antibacterial agent which could control the growth of lethal pathogenic bacteria.

REFERENCES

1. Attaie, R., Whalen J., Shahani K.M., Amer, M.A., 1987. Inhibition of growth of S. aureus during production of acidophilus yogurt. J. Food Protec. 50, 224-228.
2. Bagchi, S.N., Palod, A., Chauhan, V.S., 1990. Algicidal properties of a bloomforming blue-green alga, *Oscillatoria* sp. J Basic Microbiol. 30, 21-29.
3. Barchi, J.J., Moore, J.R.E., Patterson, G.M.L., 1984. Accutiphycin and 20, 21-Didehydroaccutiphycin, new antineoplastic agents from the cyanophyte *Oscillatoria acutissima.* J Am Chem Soc. 106, 8193-8197.
4. Bloor, S., Englan, R.R., 1991. Elucidation and optimization of the medium constituents controlling antibiotic production by the cyanobacterium *Nostoc muscorum*. Enzym. Microb. Technol. 13, 76-81.
5. Borowitzka, A.M., 1995. Microalgae as source of pharmaceuticals and biologically activecompounds. J. Appl. Phycol. 7, 3-15.
6. Borowitzka, M.A., Borowitzka, L.J., 1992. Microalgal Biotechnology. Cambridge University Press. USA.
7. Carmichael, W.W., He, J.W., Carmichael, W.W., Eschendor, J., He, Z.R., Juan, Y.M., 1988. Partial structural determination of hepatotoxic peptides from *Microcystis aeruginosa* (cyanobacterium) collecterd in ponds of central China. Toxicon. 26, 1213-1217.
8. Chetsumon, A., Fujieda, K., Hirata, K., Yagi, K., Mura, Y., 1993 Optimization of antibiotic production by the cyanobacterium *Scytonema* sp. TISTR 8208 immobilized on polyurethane foam. J. Appl. Phycol. 5, 615-622.
9. Chetsumon, A., Maeda, I., Umeda, F., Yagi, K., Mura, Y., Mizoguchi, T., 1994. Antibiotic production by the immobilized cyanobacterium, *Scytonema* sp. TISTR 8208, in a seeweed-type photobioreactor. J. Appl. Phycol. 6, 539-543.
10. Chetsumon, A., Maeda, I., Umeda, F., Yagi, K., Mura, Y., Mizoguchi, T., 1995. Continuous antibiotic production by an immobilized cyanobacterium in a seewead- type bioreactor. J. Appl. Phycol. 7, 135-139.

11. Debro L. H. and Ward H. B. (1979) Antibacterial Activity of Freshwater Green Algae. Planta Med. 36(8): 375-378.
12. de Mule, M., de Caire, G., de Cano, M., Halperin, D., 1991. Bioactive compound from *Nostoc muscorum* (Cyanobacterium). Cytobios. 66, 169-172.
13. Fish, S.A., Codd, G.A., 1994. Analysis of culture conditions controlling the yield of bioactive material produced by the thermotolerent cyanobacterium (blue green alga) Phormidium. Eur. J. Phycol. 29, 261-266.
14. Ishida, K., Matsuda, H., Murakami, M., Yamaguchi, K., 1997. Kawaguchipeptin B, an antibacterial cyclic undecapeptide from the cyanobacterium *Microcystis aeruginosa*. J. Nat. Prod. 60, 724-726.
15. Khan NH, Rahman M, Kamal Nur-E., 1988. Antibacterial activity of *Euphorbia thymifolia* Linn Indian J Med Res. 87, 395-397. Knübel, G., Larsen, L.K., Moore, R.E., Levine, I.A., Patterson, G.M.L., 1990. Cytotoxic, antiviral indolocarbazoles from a blue green alga belonging to the Nostocaceae. J. Antibiotics. XLIII, 1236-1239.
16. Kreitlow, S., Mundt, S., Lindequist, U., 1999. Cyanobacteria- A potential source of new biologically active substances. J. Biotech. 70, 61-63.
17. Noaman, N.H., Fattah, A., Khaleafa, M., Zaky, S.H., 2004. Factors affecting antimicrobial activity of *Synechococcus leopoliensis*. Microbiological Research, 159, 395-402.
18. Özdemir, G., Conk Dolay, M., Küçükakyüz, K., Pazarba?i, B., Yilmaz, M., 2001. Determining the antimicrobial activity capacity of various extracts of Spirulina platensis produced in Turkey's conditions. J Fisheries and Aquatic Sciences 1st. Algal Technology Symposium. 18/1, 161-166.
19. Rippka, R. M., 1988. ethods in Enzymology. London, Academic Press, Vol 167, pp 327.
20. Stewart, J.B., Bornemann, V., Chen, J.L., Moore, R.E., Caplan, F.R., Karuso, H., Larsen, L.K., Patterson, G.M.L., 1988. Cytotoxic, fungicidal, nucleosides from blue green algae belonging to the Scytonemataceae. J. Antibiotics. XLI, 1048-1056.
21. Vlachos, V., Critchley, A.T., von Holy, A., 1996. Establishment of a protocol for testing antimicrobial activity in southern African macroalgae. Microbios 88, 115-123.

Floral Diversity and their Conservation (2013), Editors: D.R. Khanna et al.
Published by Biotech Books.
ISBN: 978-81-7622-286-0 Pages: 125-134

14

Standing Crop Biomass and Net Primary Production of *P. Graveolens* Cutivation as Affected by Foliar Application of Different Concentration of Zinc Sulphate

Jai Laxmi Rawat* and A.K. Agrawal
Department of Botany, Govt. P.G. College, Uttarkashi (Uttarakhand) India

Present study deals with the impact of foliar application of different concentration of Zinc Sulphate (ZnSo4) on standing crop biomass and net primary production of *P. graveolens* cultivation in mid Himalayan region. Zinc Sulphate was applied as foliar spray in concentration of 10, 20, 30 and 40ppm. It was observed that foliar spray of 30ppm Zinc Sulphate significantly increased the standing crop biomass and net primary productivity of herbage. However, application of 40ppm Zinc Sulphate affects adversely. It was recommended from the present study to enhance herbage production and ultimately the essential oil of *P. graveolens* the crop should be sprayed 30ppm Zinc Sulphate in mid Himalayan environment of Uttarakhand.

Key words: Zinc Sulphate, *P.graveolens*, Crop Biomass, Foliar application

INTRODUCTION

Aromatic plant should be cultivated to save the forest and Himalayan regions and at the same time meet the rising demands of herbs to improve the livelihood of the people of mid Himalayan region. It is well known that the cultivation of food grains in the hill is not economically viable, and thus immediate diversification from agriculture to economically viable and sustainable alternatives is needed. *Pelargonium graveolens* L. Her. Ex Ait., belongs to family Geraniaceae is an important aromatic plant which gives light yellow to light brown coloured essential oil on steam distillation of fresh herb. Geranium oil is widely used in high grade perfumes. The total world production of Geranium oil is about 300 tonnes annually (Muni Ram and Gupta, 1995). The demand as well as the price of this oil in India are increasing with the passage of time. In the beginning, though its cultivation was confined to hill stations of Tamil Nadu (1402 ha) and plains of South India producing 20 tonnes in 1980 (Narayana and Prakasa Rao, 1986). Recently it has been shown that the crop can also be successfully grown in North India (Muni Ram and Gupta, 1995., Sushil Kumar, 1996).

The use of fertilizers as a form of soil management has helped to improve yields of crop, but has also brought about nutritional unbalance. Maximum yield should be the result of an optimal specific combination of nutrient concentrations. An element can affect not only another element concentration but also its critical concentration (Olsen, 1972). Zinc plays an important role for maintenance of the physical organization and activities in living cells by virtue of its function in various metabolic processes. Apart from being constituent of various enzymes, the functional role attributed to Zinc includes: Auxin metabolism, tryptophan synthatase and tryptamine metabolism, catalyst of dehydrogenase enzymes- pyridine nucleotide, alcohol, glucose-6-phosphatase and triose phosphate, phosphodiesterase and pyruvic caboxylase, carbonic anhydrase (localized in chloroplasts), super oxide dimutase. It also promotes synthesis of cytochrome C; stabilizes ribosomal fractions and photosynthesis and nitrogen metabolism etc Sommer and Lipman (1926). Heavy metal pollution in soil caused either by natural process or by human activities is the most serious environmental problem. Copper and Zinc are essential nutrients for normal growth and development of plants when they are in lower concentration. Higher concentration of Copper and Zinc are toxic to plant growth (Ganga devi *et al*., 2001). Zinc is now regarded as the key element among essential micronutrients elements. However, its essential nature was not generally accepted until Sommer and Lipman (1926). Sommer showed with experimental evidences that Zinc was indispensable for normal growth of plants. Zn and S deficiency is widespread in the soils of Northern India (Dev and Arora, 1984; Takkar *et al.*, 1975). Plant roots absorb Zinc as the ion Zn^{2+} and as component of synthetic and natural molecular complexes. Soluble Zinc salts and Zinc complexes can also enter the plant system directly through leaves.

Zinc deficiency develops characteristics symptoms. Internodes fail to lengthen and leaves fail to develop, resulting in dwarfed plants as in grasses or resetting in many tree species (Chapman, 1966). Along with shoot symptoms, old leaves of Zn deficient plants frequently exhibit a chlorotic molting called 'frenching' in citrus or as in the case of corn a white striping in the lower half of older leaves (Chapman, 1966). Plants differ in their sensitivity to Zn deficiency, with corn, cotton, and citrus being considered sensitive, and barley, oats and rye being less so (Chapman, 1966).

Considering the facts about the plant diversity, their present status, agricultural strategies, present study was undertaken to introduce Geranium (*Pelargonium graveolens*) in mid Himalayan region.

MATERIALS AND METHOD

Present study was undertaken at village Genwala near Uttarkashi. The village is located at 30^0 22' North latitude and $77^0$51' East longitudes at an elevation of 900m above mean sea level, in central Himalayan Zone of Garhwal Himalaya. The experimental site was located near a rivulet named Ranu Ki Gad, a tributary of river Bhagirathi. A land area of 334 sqm, divided in to five plots, was selected to study the growth behavior, productivity and essential oil content of crop.

GENERAL EXPERIMENTAL DESIGN

During the field studies, the crop was grown in the field in the month of February 2004. The fields were ploughed twice before crop sowing. The pre soaked and fungicides treated shoot cuttings of *Pelargonium graveolens* was sown in the month of January 2004 in five plots. Following treatments were given to the crops after two leaf stage and thereafter one month's interval for one year during 2004-2005. The data were collected after two leaf stage continuously for one year at regular interval of one month.

Plot 1	Plot 2	Plot 3	Plot 4	Plot 5
Control Plot (No treatment)	Zn 10ppm	Zn 20ppm	Zn 30ppm	Zn 40ppm

METHODOLOGY

The shoot cuttings were sown in the month of January 2004 at experiment site. The field for cultivation was prepared before planting the cuttings as proposed by Dhasmana (1984). The general experimental plans for different treatment were laid after two-leaf stage (Kumar, 1981; Dhasmana, 1984; Ambrish, 1992; Chandhok, 2005).

STANDING CROP BIOMASS

In the present study, estimation of standing crop biomass was evaluated by studying the increment in plant biomass (short-term harvest method; Odum, 1960), *i.e.*, changes in standing crop. In the application of this method it is necessary to harvest the standing crop at several times during the study period of fix time interval (Agrawal, 1989). For the measurement of above ground and underground plant biomass, ten plants from each plot were harvested at monthly interval. The difference between any two consecutive sampling dates for each species was of one month.

The above ground plant parts were cut with the help of sharp cutter in order to obtain a valid estimation of herbage. All the samples were washed several times packed in separate polythene bags and brought to the laboratory were separated in to leaves stem and root.

All the samples were dried at room temperature with blotting paper. Fresh weight of all the components of the harvested material was determined. The samples were put in hot air oven at 800c for 24 hours and weighted. The dry weight of all the samples component (weighed by electric single pan balance) were averaged and computed as kg/ha.

The net primary productivity was calculated by subtracting the values of biomass from that of subsequent growth stage. Summation of positive increments represents the net primary productivity of each component. Total sum of positive values of each part gives the values for total net productivity of the crop (Odum, 1960; Agrawal, 1989 and Bhatt, 2003)

RESULT AND DISCUSSION

Standing Crop Biomass

The minimum above ground biomass was recorded at initial stage after one month growth and amounting to 873.0 kg/ha. The value was found to increase linearly till last stage of growth in January 2005 and amounting to 422.6 Kg/ha in T1 Plot. However, the maximum values were recorded in T3 Plot amounting to 922.2, 1350.4, 1933.4, 2444.2, 2699.2, 3291.2, 3422.1, 3156.2, 3254.0, 3655.2, 3833.5, 4423.3 kg/ha.

Table 1: Above ground biomass of *Pelargonium graveolens* as affected by different treatments (kg/ha).

Treatment	Feb04	Mar04	Apr04	May04	Jun04	Jul04	Aug04	Sep04	Oct04	Nov04	Dec04	Jan05
T1	936.0	1432.8	2050.2	2460.6	2860.2	3198.6	3250.8	3322.8	3383.0	3904.2	3855.5	4347.0
T2	908.0	1432.6	2016.2	2475.6	2943.5	3144.6	3230.8	3328.8	3387.7	3906.3	3860	4345.6
T3	886.4	1452.4	2135.3	2500.4	2921.1	3350.3	3247.3	3307.9	2377.9	3821.5	3310.1	4471.5
T4	992.4	1491.0	2264.8	2783.2	2957.4	3492.4	3585.4	3345.1	3502.3	4065.4	3902.7	4587.0
T5	850.0	1382.2	2096.7	2408.8	2844.8	3318.5	3400.5	3141.4	3241.6	3741.7	3682.9	4077.8

Data presented in Table 2. reveal the results of underground biomass of *Pelargonium graveolens* in different months of growth as affected by different concentration of Zn treatments. There was a linear growth of biomass of underground parts in all the treatments. The minimum biomass was recorded at one – month interval and amounting to 63.0 ±1.1, 62.0 ±1.3, 62.4 ±1.7, 70.2 ±0.9, 60.6 ±0.7 kg/ha in T1, T2, T3, T4, T4 and T5 plot respectively. However, the maximum values were recorded in November 2004 and amounting to 396.0±11.6, 398.3±13.4, 399.4±17.9, 410.2±11.8, 320±9.7 Kg/ha respectively in T1, T2, T3, T4 and T5.

Table 2: Under ground biomass of *Pelargonium graveolens* as affected by different treatments (kg/ha)

Treatment	Feb04	Mar04	Apr04	May04	Jun04	Jul04	Aug04	Sep04	Oct04	Nov04	Dec04	Jan05
T1	63.0 ±1.1	138.6 ±2.2	313.2 ±9.6	327.6 ±3.3	259.2 ±2.6	145.8 ±2.1	163.8 ±2.1	183.6 ±2.6	245.6 ±3.6	396.0 ±4.6	93.5 ±1.2	122.4 ±4.2
T2	62.0 ±1.3	134.6 ±2.1	316.2 ±5.2	329.6 ±2.3	262.2 ±3.2	135.8 ±2.4	133.8 ±3.2	188.6 ±2.5	249.3 ±3.1	398.3 ±13.4	93.5 ±1.1	122.4 ±3.6
T3	62.4 ±1.7	144.2 ±1.2	324.1 ±6.3	346.3 ±3.1	233.1 ±6.2	139.1 ±2.1	136.0 ±3.2	185.8 ±2.4	244.9 ±3.1	399.4 ±17.9	65.9 ±2.1	160.4 ±6.6
T4	70.2 ±0.9	140.6 ±2.1	331.4 ±8.2	339.0 ±2.2	258.2 ±5.3	201.2 ±1.5	163.3 ±3.3	188.9 ±1.1	248.3 ±3.6	410.2 ±11.8	69.2 ±2.1	163.7 ±6.2
T5	60.6 ±0.7	115.0 ±3.2	311.5 ±8.2	320.2 ±3.7	244.3 ±3.6	195.3 ±1.5	156.3 ±3.4	116.2 ±1.6	230.4 ±3.5	320.5 ±9.7	59.6 ±2.2	154.2 ±5.9

Table 3: Total biomass of *Pelargonium graveolens* as affected by different treatments (kg/ha)

Treatment	Feb04	Mar04	Apr04	May04	Jun04	Jul04	Aug04	Sep04	Oct04	Nov04	Dec04	Jan05
T1	873.0	1294.2	1737.0	2133.0	2601.0	3052.8	3087.0	3139.2	3137.4	3508.2	3762.0	4224.6
T2	846.0	1298.0	1700.0	2146.0	2681.3	3008.8	3097.0	3140.2	3138.4	3508.0	3766.5	4223.2
T3	824.0	1308.2	1811.2	2154.1	2688.0	3211.2	3111.3	3122.1	2133.0	3422.1	3244.2	4311.1
T4	922.2	1350.4	1933.4	2444.2	2699.2	3291.2	3422.1	3156.2	3254.0	3655.2	3833.5	4423.3
T5	789.4	1267.2	1785.2	2088.6	2600.5	3123.2	3244.2	3025.2	3011.2	3421.2	3623.3	3923.6

Table 4: Total Above ground biomass production *Pelargonium graveolens* as affected by different treatments (kg/ha)

Treatments	Feb04	Mar04	Apr04	May04	Jun04	Jul04	Aug04	Sep04	Oct04	Nov04	Dec04	Jan05
T1	873.0	421.2	442.8	300.3	396	919.8	34.1	52.1	-	370.8	253.8	462.6
T2	846.0	452.0	402.0	468.0	446.0	862.8	88.1	43.1	-	369.6	258.5	456.7
T3	824.0	484.2	503	535.3	342.9	1057.1	-	10.7	-	128.1	-	106.9
T4	922.2	428.2	583	533.9	510.8	847.0	130.9	-	97.2	401.2	178.3	589.8
T5	789.4	477.8	518.0	255.0	303.4	1034.6	121.0	-	-	410.0	202.1	300.3

Standing crop biomass represents the long-term integration of biochemical, physiological and growth parameters of the plant. Therefore, Zn in different concentrations showed marked impact in biomass partitioning. Standing crop of biomass of the plant parts was found to increase consistently and linearly till last stage of growth in all treatments. The maximum impact was observed in T4 (Zn 30ppm) in all plant parts. It is therefore clear that Zn 30ppm enhanced the biomass favorably in favorable concentration.

Net Primary Production

Net primary production of above ground biomass of *Pelargonium graveolens* given in Table 4. No consistent trend was noted for net primary productivity of above ground biomass of *Pelargonium graveolens*.

Net primary productivity of underground plant parts of *Pelargonium graveolens* given in Table 5, indicate that minimum values were recorded at initial stage after one month of growth in all the plots and amounting to 63.0, 62.0, 62.4, 70.2, 60.6 kg/ha in T1, T2, T3, T4 and T5 plots respectively. The values of net primary productivity of underground parts of *Pelargonium graveolens* were found to increase consistently and linearly up to April where they showed their maximum values. After attaining their maximum, these values were found to decrease and showed no particular trend.

Table 5: Under ground biomass production of *Pelargonium graveolens* as affected by different treatments (kg/ha)

Treatments	Feb04	Mar04	Apr04	May04	Jun04	Jul04	Aug04	Sep04	Oct04	Nov04	Dec04	Jan05
T1	63	75.6	174.6	14.4	-	-	-	19.8	62	150.4	-	-
T2	62	72.6	181.6	13.4	-	-	18	54.8	60.7	149	-	-
T3	62.4	81.8	179.9	22.2	-	-	-	49.8	59.1	154.5	-	-
T4	70.2	70.4	190.8	7.60	-	-	-	25.6	59.4	161.9	-	-
T5	60.6	54.4	196.5	8.6999	-	-	-	21.0	114.2	90.1	-	-

Table 6: Net primary productivity of *Pelargonium graveolens* as affected by different treatments (kg/ha).

Treatments	Feb04	Mar04	Apr04	May04	Jun04	Jul04	Aug04	Sep04	Oct04	Nov04	Dec04	Jan05	Total Net Production
T1	936	496.8	617.4	314.7	396	919.8	34.1	953.9	988	370.8	253.8	462.6	6743.9
T2	908	524.6	583.6	481.4	446	862.8	88.1	950.9	1039	369.6	258.5	456.7	6969.2
T3	886.4	566	682.9	557.5	342.9	1057.1	25.0	1057.1	1057.1	128.1	11.0	106.9	6492.0
T4	992.4	498.6	773.8	541.5	510.8	847	130.9	977.9	1108.8	401.2	178.3	589.8	7551.0
T5	850	532.2	714.5	263.6	303.4	103.6	121.0	21.0	121	410	202.1	300.3	3942.7

Total net primary productivity of *Pelargonium graveolens* was found maximum in T4 Plot (7551kg/ha/year) and minimum in T5 Plot (3942.7 kg/ha/year). The growths of plants were found very well during summers, but Fusarium attack was noticed during rainy season where the growth was retarded. In winter, the growth took speed again. Treatment of Zinc at the rate of 30ppm per month significantly increased the standing crop biomass and net primary productivity of herbage. However, application of 40ppm Zinc Sulphate affects adversely. It was recommended from the present study to enhance herbage production and ultimately the essential oil of *P. graveolens* the crop should be sprayed 30ppm Zinc Sulphate in mid Himalayan environment of Uttarakhand.

ACKNOWLEDGEMENTS

Corresponding author is greatful to Department of Botany Govt. P.G. College Uttarkashi and Dr. Arun Kumar Agrawal Assistant Professor, Department of Botany for devoting his valuable time till the completion of this work.

CONCLUSION

It can be concluded from the present study that the treatment of Zinc at the rate of 30ppm per month as foliar spray was found suitable for increased biomass potential and net primary productivity as compared to control plot (no treatments were given) and other treatments. When Zn 40ppm was applied it shows negative effect.

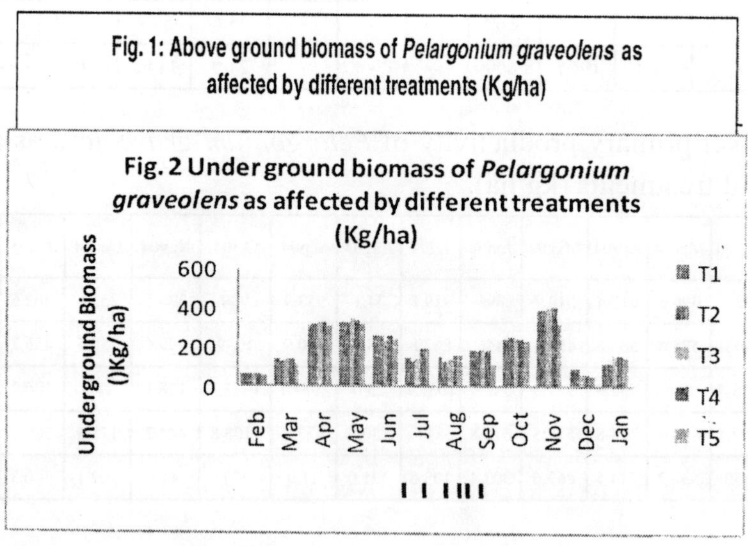

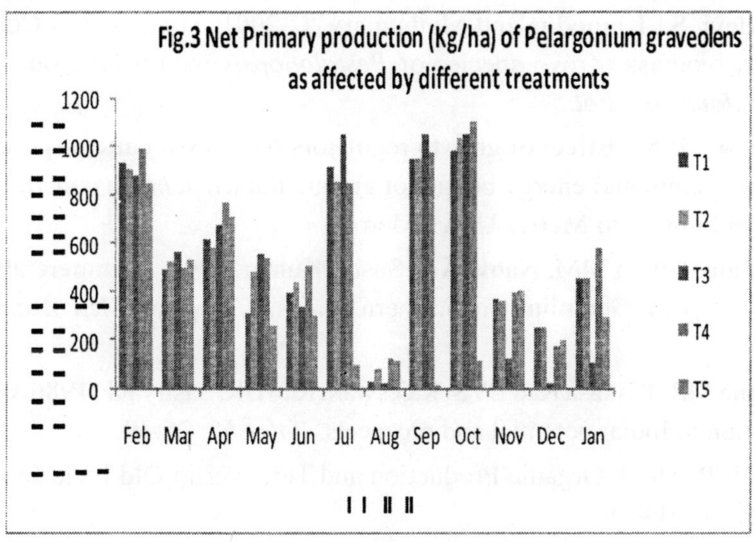

REFERENCES

1. Agrawal, A.K., 1989. Structure, Production, Energy Budget and Mineral Cycling of temperature grassland of temperate grassland of Dwarikhal (Garhwal Himalaya). D. Phill. Thesis, University of Garhwal, Srinagar Garhwal.
2. Ambrish, K.,1992. Effect of Supplement Ultraviolet-B Radiation on Growth and Composition of certain legume crops. D. Phill. Thesis, HNB University of Garhwal, Srinagar (Garhwal).
3. Bhatt, N., 2003. Effect of Enhanced UV-B Radiation on Growth relation Biomass Partitioning Mineral Cycling and transfer functions of Solanaceous plants (*Lycopersicum esculentum* & *Capsicum annum*). D. Phill. Thesis, H.N.B. University of Garhwal Srinager (Garhwal).
4. Chandhok, A., 2005. Effect of different concentrations of α-NAA and Kinetin of different Ecophysiology parameters of *Withenia somnifera* and *Solanum khasium*. D. phill. Thesis, Univ. of Garhwal, Srinagar (Garhwal).
5. Chapman, H.D., 1966. Zinc. *In* Diagnostic criteria for plants and soils. Ed. H.D. Chapman. University of California, Div. *Agric. Sci.*, Pp 484-499.
6. Dev, G. and Arora, C.L., 1984. Crop response to secondary nutrients in the northern region of India. Proc. FAI-NR Seminar held at Jaipur, India (March 30-31. 1984) Pp 135-146.
7. Dhasmana, R., 1984. Effect of growth regulators on productivity, Energy Budget and Mineral Cycling of *Medicago sativa* Linn. D. Phill. Thesis Garhwal University, Srinagar (Garhwal).

8. Gangadevi, S., Kumodha and Muthumary, J., 2001. The effect of Copper and Zinc on biomass of two species of *Pestalotiopsis* with a note on uptake of metals. *Jour. Ind. Bot.*

9. Kumar, A., 1981. Effect of growth regulators on growth pattern, productivity, mineral cycling and energy budget of ground nut (*Arachis hypogea*). D. Phill. Thesis submitted to Merrut Univ., Merrut.

10. Muni Ram, Gupta MM, Naqvi AA, Sushil Kumar, 1995. Commercially viable annual crop of Geranium in northern plains. *Curr Res Med Arom Pl* 17; 17-20.

11. Narayana MR, Prakasa Rao EVS, Rajeswara Rao BR, Sastry KP. 1986, Geranium cultivation in India; potential and prospects. *Pafai J* 8: 25-30.

12. Odum, E.P., 1960. Organic Production and Turnover in Old Field Succession. *Ecology*, 41: 34-49.

13. Olsen, S.R., 1972. Micro nutrients interactions, In: J.J. Mortvedt, P.M. Giordano and W.L. Lindsay (eds.) Micro-nutrient in Agriculture. Soil Science Society of America, Madison, WI. Pp. 243-264.

14. Sommer, A.L. and Lipman, C.B., 1926. *Plant and Physiology*, 1: 23.

15. Sushil Kumar, Muni Ram. 1996. Geranium cultivation schedulein the winter cropping season of north Indian plains. *J Med Arom Pl Sci* 18: 264-265.

16. Takkar, P.N., Mann, M.S. and Randhawa, N.S., 1975. Effect of direct and residual available Zn on yield, Zn concentration and its uptake by wheat and groundnut crops. *J. Indian Soc. Soil Sci.*, 23, 91-95.

Floral Diversity and their Conservation (2013), Editors: D.R. Khanna et al.
Published by Biotech Books.
ISBN: 978-81-7622-286-0 Pages: 135-145

15

Medicinal Plants: Ayurvedic and Homoeopathic Perspective

Veena Joshi
Associate Professor, Department of Chemistry,
HNB Garhwal University SRT Campus Badshahi Thaul

Traditional medicines are used by about 60% of the world's population. These are used for primary health care,not only in rural areas of developing nations but they are also used in the developed countries where modern medicine are pre dominantly used .In the western world the use of medicinal herbs is continuously growing, approximately 40% of the population is using herbs for medical illness due to increased incidences of adverse effects of allopathic medicine. There are about 45000 plant species in India, Eastern Himalayas, Western Ghats and Andman and Nicobar Islands are the hot spot for medicinal plants. Officially documented plants with medicinal potential are 3000 but traditional practioner use more than 6000. Seventy percent of the population in the rural India is dependent on the ayurvedic system of medicine. Most of the drugs used in modern medicine and ancient Indian medicinal system are of plant origin. Beside plants many minerals, salts and animal products are used in Ayurvedic medicines Homoeopathy originated in west, German physician Samuel Hanemann was the father of homoeopathy (1796), the homeopathic remedies are prepared by successive dilution followed by shaking forcefully. Homoeopathy uses animal, plant, mineral, and synthetic substances in its remedies. Arsenicum album (arsenic oxide), Natrum muriaticum (sodium chloride), opium (plant), and thyroidinum (thyroid hormone) are some of the homoeopathic medicines extracted from different sources.

INTRODUCTION

Ayurvedic medicine is a system of traditional medicine,native to India. In Sanskrit word **ayush** means longevity and **veda** means related to knowledge.[1] The earliest literature on Indian medical practice appeared during the vedic period in India. All the Vedas-Rig, Yajur, Sam, and Atharv have contributed to the development of Ayurveda. The *Shushrut samhita* and the *charaka samhita* are great encyclopedias of medicine compiled from various sources from the 600-BC to about 500. *Charaka samhita* has mentioned about 341 plants while *Shushrut samhita* have listed 760 medicinal plants[2].[3] They are among the foundational works of Ayurveda. A number of drugs and surgical methods are developed by ayurvedic practitioners for various ailments. Ayurveda stresses on a balance of three elemental energies or humors vata (air & space - "wind"), pitta (fire & water -"bile") and kapha (water & earth "phlegm"). According to ayurvedic medical theory, these three dosas are important for health, because when they exist in equal quantities, the body will be healthy, and when they are not in equal amounts, the body will be unhealthy. Ayurveda stresses on the use of plant-based medicines and treatments. Hundreds of plant-based medicines are employed by ayurvedic practioners .Some animal products may also be used, like milk, bones .and minerals including sulphur, arsenic, lead, copper sulphate. Some metals like gold,silver and mercury are also consumed as prescribed. Many ayurvedic herbs used for therapy have shown very promising results like turmeric and its derivative curcumin[4] are very good antioxidants. Tinspora cordifolia has been tested.for its hepato- protective nature, Salvia officinalis (Common sage) may improve Alzeheimer's patients[5]. Many plants used as **rasayana** (rejuvenation) medications are found potent antioxidants.

Homeopathy is a form of alternative medicine in which patients are treated with highly diluted preparations that are supposed to show similar symptoms in healthy persons as the patients. The basic principle of homeopathy, known as the "law of similars", is "let like be cured by like."[6] It was first stated by German physician Samuel Hanemann in 1796 .Homeopathy is a vitalist philosophy that interprets diseases and sickness as caused by disturbances in a hypothetical life- force(vital force) . Homeopathy states that the vital force has the ability to react and adapt to internal and external causes, which homeopaths refer to as the law of susceptibility. Hahnemann proposed homeopathy in reaction to the traditional western medicine at that time, which often was brutal and more harmful than helpful. He believed that by using drugs to induce symptoms, the artificial symptoms would stimulate the vital force, causing it to neutralise and expel the original disease . Homeopathic remedies are prepared by successive dilution with shaking forcefully, it uses many animal, plant, mineral, and synthetic substances in its remedies . Depending on the dilution, homeopathic remedies may not contain any pharmocologiclly active molecules (12x)[7], it is based on the belief that a substance in large amount will produce a symptoms of a specific disease will cure it in very small doses and during the process of dilution a lot of energy is released which helps in curing the ailment .

SOME MEDICINAL PLANTS USED IN AYURVEDA

1. Withania somnifera: *Withania somnifera* commonly known as *Ashwagandha* is an important plant of family Solanaceae known for its rejuvenating properties[8,9,10] it is also known as Indian Ginseng. It is a very important herb of ayurvedic indigenous medical system known for more than 3000 years .*Withania somnifera* is a very small woody shrub native to drier parts of India.. It is commonly found in India, Pakistan, Afganistan and Africa . It is about 6 feet in height and herbaceous in nature, stem bases of the plant are thickened,cylindrical and green , leaves are erect ,ovate in shape petiolate and glaberous. Roots are thickened, branched and brown in colour ,fruits are orange coloured berries enclosing many small seeds covered with green persistent calyx.

Chemical Constituents: The major constituents of *Ashwagandha* root are steroidal alkaloids and steroidal lactones of a class of compounds known as withanolides .Near about 12 alkaloids and 35 withanolides have been isolated from the plant so far. The whole plant is used for the extraction of compounds. From leaves 9 steroidal lactones of withanolide series from E-

M are extracted .Stem bark contains some withanolides while root contains few alkaloids ,flavonolides and free amino acids.

Medicinal Properties: *W.somnifera* inhibits growth in central nervous system, lung, colon and breast cell illness[11]. Extract of *Withania somnifera* was found helpful in preventing tumour growth in cancer patients. Studies show that ashwagandha possesses anti-inflammatory, antistress, antioxidant, immunomodulatory, hemopoetic, and rejuvenating properties. It also appears to exert a positive influence on the endocrine, cardiopulmonary, and central nervous systems. Found effective in the treatment of osteo-arthritis inflammation. According to the ayurvedic practioners in ashwagandha Rasa is tikta, Guna –snigdha , Veerya- ushna and Vipaka is madhura.It is helpful in kapha and and vata dosas,and very effective in controlling blood sugar and cholesterol level .It improves over all health so given as rejuvenating drug to aged persons.

2. Tinospora cordifolia: *Tinospora cordifolia* belongs to the family Menispermaceae and is commonly known as giloiya, guduchi or amruta [12], the plant is commonly known as Giloya, which is a Hindu mythological term that refers to the heavenly elixir that have saved celestial beings from old age and kept them eternally young. It is used in ayurvedic medicine from a long time (the traditional medicine of India). *Tinospora cordifolia* is a large, glabrous, deciduous climbing shrub[13]. It is distributed throughout tropical Indian subcontinent, China, Pakistan and Srilanka, ascending to an altitude of 300 m. The stem of *Tinospora cordifolia* is succulent with long hanging aerial roots from its branches. The bark is creamy white to grey,leaves are membranous and heart shaped, flowers are small and yellowish green in colour. Male flowers are in cluster and female flowers are solitary. Fruits are red,ovoid in shape,succulent and single seeded drupe.

Chemical Constituents: From the stem of *Tinospora cordifolia*[14] alkaloid berberin, tinosporin and palmitin are isolated while from roots,tinosporin and palmatine are isolated. From the whole plant, the diterpenoiddal lactone tinosporide and tinosporon are obtained. Beside these compounds giloin, gilonin and tinosporic acid are also isolated from the whole plant of *Tinospora cordifolia* .

Medicinal Properties: *Tinospora cordifolia* or giloya is widely used medicine in ayurvedic system of medicine for its general tonic, antiperiodic, anti-spasmodic, anti-inflammatory, anti-arthritic, anti-allergic and anti-diabetic properties[15]. The plant is used in ayurvedic, "Rasayanas" to improve the immune system and the body resistance against infections. The root of this plant is known for its ant-istress, anti-leprotic and anti-malarial activities The stem is bitter, stomachic, diuretic,stimulates bile secretion, causes constipation, allays thirst, enriches the blood and cures jaundice. The extract of its stem is useful in skin problems. The root and stem of *T. cordifolia* are prescribed in combination with other drugs as an anti-dote to snake bite and scorpion. Dry barks of *T. cordifolia* has anti-spasmodic antipyretic, anti-allergic, anti-inflammatory [16,17] and anti-leprotic properties. According to ayurvedic practioners Rasa of the plant are –tikta, Virrya is heating and Vipaka is madhura . It is widely used for urinary complaints and rheumatism.

3. *Calotropis gigentica*: *Calotropis gigentica* is a plant of family Asclepidaceae commonly called milk-weed or madder .It is found through out the India in dry waste lands upto a height of 900m. Madar is a large hard much branched shrub, 3-4 feet in height. The whole plant is used as medicine, it is covered with soft white wool. Leaves of the plant are opposite, sub sessile and ovate in shape.Flowers are beautiful purple and white in colour,fruits are fleshy and green in colour .

Chemical Constituents: Latex of the plant *Calotropis gigentica* contains the cardiac glycosides, a complex mixture of chemicals, some of which are steroidal heart poisons known as "cardiac aglycones. *Calotropis* contains some glycosides known as calotropin, calotoxin, calactin. uscharidin and voruscharin are the sugars with nitrogen and sulphur in their structures .Lupeol is isolated from latex. Quercetin-3-rutinoside is identified in the roots, stem, leaves, flowers and latex.

Medicinal Properties: The whole plant is used as medicine, its latex is used for the skin problems. In Ayurveda the madar plant is used for asthma ,bronchitis ,dyspepsia, and swelling [18,19,20]. Heated leaves are applied on the painful and swollen joints, it relieves the pain due to arthritis. Powdered root bark of the plant is given to the patient suffering from jaundice, asthama and bronchitis. It is used as febrifuge, anthelmintic , expectorant and antidote to snake bite.Its latex is used to induce abortion by folk people. According to Ayurveda, *Calotropis* is used in Kapha and vata dosa, its Rasa is tikta, Virrya is ushna, Vipaka is katu and Guna is snigdha. *Calotropis* is also used as an important homeopathic drug.

4. Andrographis paniculata: *Andrographis paniculata* **is a shrub of family Acanthaceae. It** grows in most places in India, including the plains and hilly areas up to 500 m, which accounts for its wide use. Native populations of *A. paniculata* are spread throughout south India and Sri Lanka which perhaps represent the centre of origin and diversity of the species. Since the time immemorial *Andrographis paniculata or* kalmegh is a popular Ayervedic and Chinese house hold remedy for the common cold, digestive issues and many other illness [21,22]. Kalmegh grows erect to a height of 30- 100 cm in height with glabrous leaves and white flowers having purple spots on the petals . Flowers are solitary and small in size, fruit is 2 cm long with numerous yellow brown seeds.

Chemical Constituents: *Andrographis paniculata* is a plant is known to possess a variety of pharmacological activities. Andrographolide bitter water–soluble bicyclic diterpenoid lactone is the major constituent extracted from the leaves, beside this compound andrographine and many flavanoids are also extracted from the root of the plant. .Neoandrographolide and Oxyandrographolide from whole plant have also been isolated

Medicinal Properties: *A. peniculata* is used in traditional ayurvedic and siddha[23] systems of medicine as well as in tribal medicine in India and some other countries for multiple clinical applications. Leaves of the herb *Andrographis paniculata* are used for ailments ranging from poor digestion to hepatitis. In the Chinese medical tradition, the plant has been used to treat everything from gastrointestinal complaints to throat infections.The plant extract exhibits anti typhoid, anti diabetic and antifungal activities. Kalmegh is also reported to possess , antibiotic, antimalarial, antithrombogenic, antiinflammatory,and anti hepato toxic, anti-snake venom, antipyretic and anti cancer drug .It is used to regulate high blood pressure but it is restricted for pregnant women. *Andrographis* balances pitta and kapha dosas , its rasa is tikta guna is ruksha, veerya is ushna and vipaka is katu .

Some medicinal plants used in Homeopathy: Homeopathy is a totally different therapy than ayurveda and allopathy, in it drug is decided by studying the symptoms of the patient . Each patient is treated by different medicines for the same ailments according to their symptoms. It is a therapy of like cures like.

5. ***Pulsetilla nigricans*****:** The genus *Pulsatilla* contains about 33 species of herbaceous perennials native to North America ,Europe and Asia Its common name is pasque flower or the meadow anemone .It is a member of family Ranunculaceae, widely used in homeopathic medicines[24,25]. The petals are deeply cut, bell-shaped, dark purple flowers and are valued for ornamental purpose, leaves are finely dissected. According to a Roman myth, the pasque flower has its origin in the tears of goddess Venus and, hence, people in ancient times used this flower to treat tearfulness. *Pulsatilla* is highly toxic, and produces cardiogenic toxins and oxytoxins. Excess use can lead to diahrroea ,vomiting, convulsions and coma.

Therapautic use of Pulsetilla: It has been found that *Pulsatilla* works best for those people who have a sweet and gentle nature. It is pre-eminently a female remedy, especially for mild, gentle, yielding disposition. Sad, thirstless, crying readily; weeps when talking. The patient seeks the open air; always feels better in open air . In 1805 Hanemann proved that pulsailla is an effective remedy for an assortment of disorders that may vary from cold and cough to gynaecological problems like delayed ,scanty and painful menstruation

6. *Bryonia alba*: *Bryonia alba* is a plant of family cucurbitaceae , it grows in Europe, North Africa and South Asia . It is a perennial climbing herb with palmately lobed leaves, flowers are white in colour and grows in auxillary clusters.The whole plant is very poisonous and fatal. It contains highly toxic compound cucurbitacin ,having antitumour and anti microbial activity.

Therapeutic use of Bryonia: Bryonia patients are emotionally, bodily and mentally **dried up[26, 27]. He wants to be left alone, undisturbed, while at the same time constantly needing great quantities of water to balance his dryness.** The Bryonia individual are most often a male. The sensation of **dryness of the mucous membranes** is most frequently reported. Symptoms of bryonia patients are dried mucous membranes ,very few or scanty expectorations, thirsty, urine very dark and scanty, stiching and tearing pains, increasing by motion .Bryonia is given to the patients suffering from constipation, stiffness of joints due to rheumatism, headache, bronchitis ,pneumonia and in measles.

7. *Aconitum napellus*: *Aconite* is a herbaceous perennial 2-5 feet in height, familiarly known as Monk's Hood, it bears beautiful hooded violet flowers that appear in mid-summer and grows from Himalayas to Europe. It is a member of family Ranunculaceae The plant is known for its poisonous alkaloids, benzaconine and aconine, found in the highest degree of concentration in the roots. Historically the poison from this plant was used as an arrow toxin. This plant was first used as a homeopathic remedy in 1805 by the founder of homeopathy, Samuel Hahnemann. This remedy is especially useful for symptoms that come on suddenly after a fright [28,29] or exposure to cold. Physical symptoms that develop as a result of shock, especially coughs, colds and fevers, respond well to this remedy.

Therapeutic use of Aconite: *Aconite* is a staple remaedy in any homeopathic first aid kit. It is useful for extreme fear and anxiety that comes on suddenly. Fear of death ,fear of croud , going out. Patient is always in great anxiety and restless,does not want to be touched, aconite patient is over sensitive to noise and can not bear music .Complaints from exposure to cold and hot weather, feels better in open air but worse in warm room, pains are intolerable, always aggrevated at night. *Aconite* is a remedy to the patients of anxiety, rheumatism, cardiac patient, and fever due to measeles and typhoid, it relieves pain and induces sweating.It is a general remedy for gastro- intestinal troubles. It is also used in ayurvedic medicine.

8. *Berberis vulgaris*: *Berberis* is a plant of family Berberidaceae an ever green shrub, 1-5m tall with thorny shoots, native to the temperate and subtropical region of Europe, Asia,Africa, and America. Species diversity is greatest in South America, Africa and Asia; Europe has a few species. *Berberis vulgaris* known as "barberry" is a thorny shrub with yellow flowers, small red fruits. Leaves are of two types some are thorny while some are leathery with spines on their edges. Flowers are beautiful and yellow in colour. Stem is brown from outside but yellow inside.Barberry fruits contain glucose, fructose, malic acid, pectine and vitamin C.In some countries fruits are used to prepare jam and jellies

The herb contains a number of active components used as haemostatic, diuretic, vasodilator, hypeotensive, antibacterial and anti-inflammatory agent. Root bark of the plant is known for containing Berberine,[30,31,32] agent which stimulates digestion and reduces the gastrointestinal pains, it is also used as antidiabetic and anti tumor agent. It is also known that this substance toughens the immune system.The bark contains a large number of alkaloids like berberine, berbamine, and tannins.

Therapautic use of Berberis: *Berberis* is used in bruised pain ,with numbness in the region of kidney. *Berberis vulgaris* patients feel stiffness and lameness with painful pressure in lumbar and renal region and chill along the spine. *Berberis* is prescribed to the patients suffering from kidney and urinary problems, gallstone pain ,burning pain in the joints and scanty menses. *Berberis* is also used in ayurveda for the liver problems.

S.No.	Botanical Name	Family	Chemical Constituents	Used for
1.	*Withania somnifera*	Solanaceae	Withanolides	Osteo- arthritis, depression
2.	*Tinospora cordifolia*	Menispermaceae	Tinosporin, tinosporic acid, gilonin	Diabetes, malaria, jaundice
3.	*Calotropis gigentica*	Asclepidaceae	Calotropin, calotoxin, lupeol from latex	Asthma, bronchitis, fever, jaundice
4.	*Andrographis paniculata*	Acanthaceae	Andrographolide, andrographine	Typhoid, cancer blood pressure
5.	*Pulsetilla nigricans*	Ranunculaceae	Anemonin, isoanemonin ($C_{10}H_8O_4$)	Fever, mumps, sore -throat
6.	*Bryonia alba*	Cucurbitaceae	Cucurbitacin, bryonin,	Synovial inflammation, pneumonia and measles
7.	*Aconitum napellus*	Ranunculaceae	Aconitin, isoaconitin, aconitic acid	Stress, anxiety fever due to typhoid ,measeles.
8.	*Berberis vulgaris*	Berberidaceae	Berberin	Liver and kidney problems

ACKNOWLEDGEMENTS

Author is highly grateful to Dr Dimple Bhandari, Dr J. P.Nautiyal Dr Bhatt and Dr R P Joshi for their valuable suggestions.

REFERENCES

1. en.wikipedia .org
2. Revitalizing Indian systems of herbal medicine by the National Medicinal Board through institutional networking and capacity, *Current science*,. **93**, (6), 2007
3. Kala, C.P., Indigenous uses, population density, and conservation of threatened medicinal plants in protected areas of the Indian Himalayas.*Conserv. Biol*, **19**, 368–378 .2005

4. Aggarwal, B.B., Sundaram, C., Malani, N., Ichikawa, H. "Curcumin: the Indian Solid Gold". *Advances in Experimental Medicine and Biology*. 595,1-75. 2007.

5. en.wikipedia.org.

6. Ernst, E , Asystematic review of systematic reviews of homeopathy, *British Journal of Clinical pharmacology*, 54 (6): 577–82, (2002),

7. Lakshmi Chandra Mishra et.al.*Alternative Medicine Review*, 5 (4), 334-346 2000.

8. Dhuley JN. Effect of *Ashwagandha* on lipid peroxidation in stress-induced animals. *J.Ethnopharmacol*; 60, 173-178,1998

9. Indigenous Ethnomedicinal plants, Trivedi P.C.Pointer publisher, Jaipur, 271-284, 2009.

10. Malhotra CL,Das P, Dhalla MS,and Prasad K, Studies on *Withania somnifera* Dunal Part IV,The effect of total alkaloids on the cardiovascular system and respiration.*Ind. .J.Med.Res* 49,448-460.1961. ,

11. Kuttan,G.,Use of *Withania somnifera* Dunal as an adjuvant during radiation therapy, *Indian J. Exp.Biol.*,34,354 -856,1996.

12. Mahendra kumar shastri *Viradh Dravya Gunadarsh* , 1st edn. Ayurvedic and Tibbi academy Uttar Pradesh ,Lucknow ,1978.

13. Rajlakshmi M,Eliza JPriya CE, Nirmala A, Daisy P. Antidiabetic properties of *Tinospora* stem extract on streptozotocin induced diabetic rats, *Afri. J. Pharm. Pharmco.*, 3(5),171-180,2009.

14. Singh S.S., Pandey SC, Shrivastav S,Gupta VS, Patro B,Ghosh AC, Chemistry and medicinal properties of *Tinospora cordifolia, Indian J. Pharmocology,* 35,83-91,2003

15. Kumar S, Verma NS, Pande D,Srivastava PS. In vitro regeneration and screening of berberine in *Tinospora cordifolia. J. Med. Arom. Plant Sci.* 22:61, 2000.

16. Bhatt RK, Sabata BK. Furanoid diterpene glucoside *fromTinospora cordifolia. Phytochemistry;*28,2419-22,1998.

17. Rao E V, Rao MV, Studies on the polysaccharide preparation (Guduchi Satwa) derived from *Tinospora Cordifolia .Indian J.pharm.Sci,*, 43,103-6, 1981.

18. Nalwaya N, Pokhrana G,DebL,Jain N K,Wound healing activity of latex of *Calotropis gigentica , International J. of Pharmacy and pharmaceutical Sci,* 1(1),176-181,2009.

19. Shilpkar P,Shah M,Chaudhary D R, An Alternative Use of *Calotropis gigentica* Biomethnation , *Current Science*, 92(4),2007

20. Manoyuan W,Wenli M,Yuanyuan D,Shenglan L,ZhumianW, Haofu D, Cytotoxic cardinolides from the roots of *Calotropis gigentica, Modern Pharmaceutical Research*, 1(2),4-9,2008.

21. Shukla PK, Chaubey O P, *Threatened Wild Medicinal Plants Assessment*, Conservation and management, Anmol Publication ,New Delhi,1st edn.116, 2007.

22. Madav. H.C,T. Tripathi, and S.K. Mishra. Analgesic, antipyretic, and anti ulcerogenic positive side effects of *andrographolide. Indian J. Pharm. Sci.* 57 (3)121-25,1995.

23. Chopra RN, Chopra IC,Handa KL, *Indigenous Drugs of India*, 2nd edn. Academic Publishers, New Delhi,1982.

24. Yaprak A E, Koroklu S T, Ketenoglu A O, A synopsis of Genus *Pulsetilla (Ranunculaceae)* in Turkey, *T urk J. Bot*, 35, 351-355, 2011.

25. Goel S.and kumar S., Anti Anxiety Activity Studies On Various extracts of *Pulsetilla nigaricans, Int.J.Pharmaceuticals and Drug Research,*2(4),291-293, 2010.

26. Farington E.A., *Clinical Materia Medica*, B.Jain Pulishers New Delhi, 1975.

27. Allen H C, *Allen's Keynoes*, B Jain Publishers, 10th edn., 2000.

28. Nautiyal BP, Prakash V, Bahuguna R, Mathani U, Bisht H, Nautiyal MC, Population Study For Monitering the Status of Rarity of three *Aconite* species in Garhwal Himalayas, *Tropical Ecology*, 43(2), 297-303, 2002.

29. Davidson JR, Morrison RM , Shore J ,Davidson RT, Bedyan G, Homeopathic treatment of Depression and Anxiety, *Alternative Therapies,* 3 (1), 46-49, 1997.

30. Fatehi M, Saleh TM, Hassanaba ZF, Farrokhfa K, Jafarzadeh M ,Davodi S, Pharmacological study on *Berberis vulgaris* Fruit extract, *J. Ethnopharmacology,*102(1), 46-52 , 2005.

31. Meliani N., Amine M E, Allali H, Tabti B, Hypoglycemic Effect of *Burberis vulgaris* in normal and Streptozotocin induced Diabetic rats, *Asian Pacific J. Tropical Biomedicine* ,468-471, 2011.

32. Ivanovska N, Phillipov S, Study on the anti-inflamatory action of *Berberis Vulgaris* root extract alkaloid fraction and pure alkaloid. Inte. J. Immunopharmocology.18 (10), 553-561, 1996.

Floral Diversity and their Conservation (2013), Editors: D.R. Khanna et al.
Published by Biotech Books.
ISBN: 978-81-7622-286-0 Pages: 147-156

16

Distributional Patterns of Freshwater Diatoms in the Highland Rivers of Indian Subcontinent

JYOTI VERMA AND PRAKASH NAUTIYAL
Aquatic Biodiversity Unit, Department of Zoology, H.N.B. Garhwal University, Srinagar, Uttarakhand, India

Fifty one, fifty and forty three genera are recorded from the West Himalaya (Garhwal region), Central Highlands and Western Ghats, respectively. Thirty six genera are common to these three biogeographic regions of the Indian subcontinent. Currently, the Indo-Gangetic Plains separate the Himalaya and the Central Highlands but are connected by the Gangetic drainage. The Western Ghats and the Central Highlands are connected like 'elbow' at northern and western extremity, respectively and extend to south and east located perpendicular to each other. These two regions constitute the Deccan Peninsula the oldest part of the Indian subcontinent, a historical reason for high similarity between these two biogeographic regions. However, similarity with the Himalaya (part of the sub continent) created historically by recent upheavals, other geological activities in the Deccan and glaciations and current geography account for this high similarity.

Preliminary investigations show presence of fifty eight genera from three biogeographic regions the Central Highland, West Himalaya and

Western Ghats of the Indian subcontinent. Fifty one taxa were recorded from the Himalaya (Garhwal region), fifty from Central Highlands and forty three from Western Ghats. Thirty six genera were common to these three regions. Historical rather than geographical factors appear to have played a greater role, due to movement of landmasses and upheavals, which not only led to changes in the drainage patterns but also climatic conditions. The recently deglaciated high altitude locations in the Himalaya have not been geologically stable while the Central Highlands and the Western Ghats in the peninsula has been geologically stable.

Key words: Diatom, Highland Rivers, India

INTRODUCTION

Many diatom species are known to have world-wide distributions; others appear limited to certain climatic zones or geographical regions, or are endemic to particular water bodies. Most of the factors found to be important for distribution of benthic river diatoms like water chemistry (pH, ionic strength and nutrient concentrations), substrate, current velocity, light and grazing (Patrick and Reimer, 1966; Round, 1981) depend strongly on climate, geology, topography, land-use and other landscape characteristics, and hence are similar within ecological regions defined by these characteristics (Stevenson 1997). The apparently global distributions of unicellular organisms have been suggested to result from huge population sizes and efficient passive dispersal over large areas, facilitated by small cell size. Kociolek and Spaulding (2000) argue that the proportion of geographically restricted diatom species is much higher than formerly thought, and because of that, the importance of geographical factors in explaining patterns in diatom flora has been underestimated. This is based on the knowledge of diatom distribution across the northern hemisphere, primarily from the continents of Europe and North America. However, there is still ignorance about the distribution of diatoms in the Indian subcontinent, especially in view of its past geology; the drift which accounts for its present position in the globe and collision with the Asian plate leading to the Himalayan orogeny. Presently, the terrain consists of plateau, plains and mountains.

Though the climate is governed by monsoon, both hot and cold wet to arid conditions prevail across the subcontinent resulting in varied ecoregions. This study examines distribution of the diatom genera, especially the taxa specific and those common to the West Himalaya, Central Highlands, and Western Ghats ecoregions, in the northern, central and southern regions of India, respectively. The study also examines the possible role of geographic and historical factors to explain the current state of distribution in these biogeographic regions.

STUDY-AREA

The study is based on diatom collections from the Chalakudy River in Western Ghats, the Ken, Paisuni and Tons in Central Highland, Yamuna, Alaknanda, Bhagirathi, Mandakini and Ganga in the West Himalaya. The geographical location of the the study area presented in Table 1.

Table 1: Geographical coordinates of the Rivers of North-Western Himalaya, Central Highland and Western Ghats.

Region	Rivers	Latitude (N)	Longitude (E)	Elevation (m)
North-Western Himalaya	Yamuna, Bhagirathi, Ganga, Alaknanda	30°8′15″ - 31°45′2″N	78°10′29″- 78°48′32″E	375-2200m
Central Highland	Ganga, Yamuna	23°59′28.92″- 25°46′15.49″N	80°0′41.16″ -80°54′26.34″E	86-365m
Western Ghats	Chalakudy River	10°05′-10°35′N	76°15′-76°55′E	0-1250m

MATERIALS & METHODS

Rationale for sampling benthic (epilithic) diatoms: Since distributional patterns of species are a function of dispersal mechanism, environmental tolerances and historical factors (Carter and Watts 1981), the likelihood of a species being present throughout a large geographical area is low. To analyze the distributional patterns, the samples are taken on the assumption that difference in distributional patterns reflects real difference among the sites and not seasonal trends (Corkum 1989). Hence, one time intensive sampling during stable flow (November to May) was deemed suitable for present study as the rivers are in floods during monsoon (mid June to mid

September) when flora is poorly represented on the substrate. Besides, the purpose was to obtain a composite sample to explore the total flora and diversity, not the seasonal trends.

Sampling: Diatom samples were collected by scraping an area of 3 x 3 cm^2 from the cobble surface and preserved in 4% formalin. Samples were treated with hydrochloric acid- peroxide, washed repeatedly and mounted in Naphrax and stored at Aquatic Biodiversity Unit. Each sample was examined under brightfield in a BX-40 Olympus microscope (x1500 oil immersion) for its flora using standard literature. The diatom genera were identified using standard literature (Hustedt and Jensen 1985; Krammer and Lange-Bertalot 1986-1991; Krammer 2002; 2003; Lange-Bertalot 2001; Werum and Lange Bertalot 2004; Metzeltin *et al.*. 2005).

RESULTS

Highland Flora

Fifty eight genera occur in three biogeographic regions of India (Table 2). Of these 51 genera occur in the West Himalaya, 50 genera in the Central Highland and 43 genera in the Western Ghats. Thirty six genera are common to all three regions, of which 42 genera are common to the Vindhya and Himalaya, 39 are common to the west Himalaya and Western Ghats while only 38 are intriguingly common to closely located Vindhya and Western Ghats. Nine genera (*Ceratoneis* Grunow, *Encyonopsis* Krammer, *Reimeria* Kociolek & Stoermer, *Rhoicosphenia* Grunow, *Melosira* Agardh, *Meridion* Agardh, *Fragilariforma* Williams & Round, *Peronia* Brèbisson & Arnott, *Nupela* Vyverman & Compère*)* are specific to the Himalaya while five (*Anomoeoneis* Pfitzer, *Gomphocymbelopsis* Krammer, *Mastogloia* Thwaites, *Aneumastus* Mann & Stickle, *Scoliopleura* Grunow) are specific to the Vindhya. Almost all these genera have either one or few species in respective ecoregions. Further investigations on other parts of the Western Ghats may yield some specific and endemic genera.

DISCUSSION

There is considerable similarity (Sorensen) among these regions; 0.83, 0.82 and 0.81 among Central Highland -West Himalaya, West Himalaya-Western Ghats and Central Highland -Western Ghats, respectively. Nautiyal and Singh (2009) observed similar distribution for freshwater fish fauna from these biogeographic regions *i.e.* low similarity among the Central Highlands and Western Ghats. It appears logical that Central Highland -West Himalaya are most similar as they are close to each other in past and recent times. The Central Highland and Western Ghats do not exhibit higher similarity in genera-richness though they lie close to each other at their respective western and northern extremities in 'elbow' like fashion. Historically, the Central Highlands and Western Ghats constitute the Deccan Peninsula the oldest (geological age) part of the Indian subcontinent another strong reason for high similarity. Lack of connection between their drainages (that flow in opposite directions), may facilitate speciation through isolation only in recent times. On the contrary, relatively higher similarity among the West Himalaya and Western Ghats is intriguing. These regions are separated by the Gangetic Plains and Central Highlands. They differ in monsoon patterns and ambient temperatures (Western Ghats, tropical latitudes; subtropical to temperate and tundra latitudes in the West Himalaya).

In the current geographic perspective, the mountain terrain is the only geographical similarity among the West Himalaya and Western Ghats. This builds up an interesting case about the distributional patterns for diatoms in the Indian subcontinent and look for possible reasons that explain the present levels of richness among the above said biogeographical regions. The, geographical factors such as terrain and climate show enough variation to support potential differences in the flora and hence genera-richness in these ecoregions. Despite these differences high similarity across the Indian subcontinent and relative difference among the biogeographic regions are attributable to historical (geological) past of the subcontinent.

The Himalayan orogeny and consequent events are other convincing reasons for looking at the geological past of the subcontinent, because Himalaya is a landmass (extra Peninsula) that has been added to the subcontinent after the Indian Plate began to slide under the Asian Plate. The Himalayan streams and rivers were virgin areas for colonization by the biota from adjacent or distant regions. Even the Western Ghats that are of recent origin in the period similar to the Himalaya were also colonised. Only the Central Highlands are ancient (Pre Cambrian). Obviously, the Western Ghats and Himalaya were colonized by elements from adjoining areas *i.e.* the old Peninsula. The genera were distributed through drainages in the then newer habitats, the Western Ghats and Himalaya. Over a period of time the regions became isolated from adjoining areas. The Himalaya became isolated from the Central Highlands as the Himalayan rivers eroded the mountains and deposited the alluvium to form the Gangetic Plains. However, the northern part of the Central Highlands *i.e.* the Vindhya drain into the Ganga river system and thus are connected till recent times.

The present similarity indicates that certain geological mechanisms led to exchange of elements over a large period to recent times after the Himalayan orogeny, despite the presence of physical barriers in the form of mountains in these biogeographic regions. Glaciations appear to be the major geological agents that facilitated these exchanges.

POSSIBLE HISTORICAL DISTRIBUTION PATTERNS AND MECHANISMS

The diatom distribution in the Himalaya, the Central Highlands and the Western Ghats may have taken the following course after Himalayan orogeny; a) initial invasion of the Himalayan rivers and the Western Ghats by peninsular elements, b) speciation in respective regions prior to the glaciations, c) downward movement of floral elements into the Peninsula through the Himalayan drainages during glaciations, d) retreat during inter-glaciations, e) the similarity may have decreased in post glaciations, as the climate became hot and taxa unable to withstand desiccation may have perished and those which modified or adapted to sub tropical climate became new species, varieties and forms, f) similar exchanges, though to a lesser degree are expected among Central Highland and the Western Ghats, where the historical factors like Himalayan orogeny have no role, while the glaciations are likely to influence southward movement of the elements.

Besides, glaciations there seem to be no geological events in the peninsula that are likely to influence diatom floral exchanges because the Peninsula has been geologically stable for over long periods except volcanic activity that may have reduced similarity among these regions. Here, the existing similarity among these two regions may be due to passive dispersal, disturbance, chance introductions as environmental factors differ among the Central Highlands (120-170 mm) and Western Ghats (200 mm) by virtue of climate.

Table 2: Distribution of diatom genera in three highland regions of the Indian subcontinent.

Genera	H	CH	WG
1. *Cyclotella* Kützing	+	+	+
2. *Diatoma* Bory	+	+	+
3. *Fragilaria* Lyngbye	+	+	+
4. *Staurosira* Ehrenberg	+	+	+
5. *Synedra* Ehrenberg	+	+	+
6. *Eunotia* Ehrenberg	+	+	+
7. *Achnanthes* Bory	+	+	+
8. *Achnanthidium* Zhakovschikovii	+	+	+
9. *Planothidium* Round & Bukhtiyarova	+	+	+
10. *Cocconeis* Ehrenberg	+	+	+
11. *Amphora* Ehrenberg	+	+	+
12. *Brachysira* Brèbisson	+	+	+
13. *Caloneis* Cleve	+	+	+
14. *Cymbella* Agardh	+	+	+
15. *Cymbopleura* Krammer	+	+	+
16. *Encyonema* Kützing	+	+	+
17. *Frustulia* Rabenhorst	+	+	+
18. *Gyrosigma* Hassall	+	+	+
19. *Gomphonema* Ehrenberg	+	+	+
20. *Navicula* Bory	+	+	+
21. *Navicula sensu lato*	+	+	+
22. *Adlafia* Moser	+	+	+
23. *Luticola* Mann	+	+	+
24. *Geissleria* Lange-Bertalot & Metzeltin	+	+	+
25. *Sellaphora* Mereschowsky	+	+	+
26. *Pinnularia* Ehrenberg	+	+	+
27. *Stauroneis* Ehrenberg	+	+	+
28. *Hantzschia* Grunow	+	+	+
29. *Nitzschia* Hassall	+	+	+
30. *Surirella* Turpin	+	+	+
31. *Cymatopleura* W.Smith	+	+	+

contd...

Genera	H	CH	WC
32. *Epithemia* Brèbisson	+	+	+
33. *Rhopalodia* O.Müller	+	+	+
34. *Neidium* Pfitzer	+	+	+
35. *Amphipleura* Kützing	+	+	
36. *Diploneis* Ehrenberg	+	+	
37. *Craticula* Grunow			
38. *Fallacia* Stickle & D.G.Mann	+	+	
39. *Hippodonta* Lange-Bertalot	+	+	
40. *Placoneis* Mereschkowsky	+	+	
41. *Denticula* Ehrenberg	+	+	
42. *Tabellaria* Ehrenberg	+	+	
43. *Aulacoseira* Thwaites		+	+
44. *Diadesmis* Kützing		+	+
45. *Bacillaria* Gmelin		+	+
46. *Ceratoneis* Grunow	+		+
47. *Encyonopsis* Krammer	+		+
48. *Reimeria* Kociolek & Stoermer	+		+
49. *Rhoicosphenia* Grunow	+		+
50. *Melosira* Agardh	+		
51. *Meridion* Agardh	+		
52. *Fragilariforma* D.M.Williams & Round	+		
53. *Peronia* Brèbisson & Arnott	+		
54. *Nupela* Vyverman & Compère	+		
55. *Anomoeoneis* Pfitzer		+	
56. *Gomphocymbelopsis* Krammer		+	
57. *Mastogloia* Thwaites		+	
58. *Aneumastus* Mann & Stickle		+	
59. *Scoliopleura* Grunow		+	

Abbreviations: (Used in Table 2)

H: Himalaya, CH: Central Highland, WG: Western Ghats

Prakash Nautiyal: Aquatic Biodiversity Unit, Department of Zoology, H.N.B. Garhwal University, Srinagar, 246174, Uttarakhand, INDIA, lotic.biodiversity@gmail.com, +0919412987878

Jyoti Verma: Aquatic Biodiversity Unit, Department of Zoology, H.N.B. Garhwal University, Srinagar, 246174, Uttarakhand, INDIA, dr.jyotiverma@ymail.com, +0919452789500

CONCLUSION

Historical rather than geographical factors appear to have played a greater role, due to movement of landmasses and upheavals, which led to changes in the drainage patterns and later in the climatic conditions. Further, creation of the Himalaya effected glaciations when the global climate was cooling. The recently de-glaciated high altitude locations in the Himalaya have not been geologically stable while the Central Highlands and the Western Ghats in the Peninsula has been geologically stable, a factor of consequence to diatom distribution in the Indian subcontinent.

ACKNOWLEDGEMENTS

The academic support by the Head, Department of Zoology, H.N.B. Garhwal University is duly acknowledged.

REFERENCES

1. Carter, J.R. and Watts, A.E.B., A taxonomic study of diatoms from standing fresh waters in Shetland. *Nova Hedwigia*, 1981; 33: 513-628.
2. Corkum, L.D., Patterns of benthic invertebrate assemblages in rivers of north-western North-America. *Freshwater Bio.*, 1989; 21: 191-205.
3. Hustedt, F. and Jensen, N.G., Die Kieseialgen Deutschlands, Oesterrichs und der Schweiz Bd. 7, Teil 2. Translated by N.G. Jensen as *The Pennate Diatoms*. Koeltz Scientific Books, Koenigstein. 1985.
4. Kociolek, J.P. and Spaulding, S.A., Freshwater diatom biogeography. *Nova Hedw.,* 2000; 71(22): 32-41.
5. Krammer, K., Diatoms of Europe - Vol 3, *The genus Cymbella*, A. R. G. Gantner Verlag K. G., 2002.
6. Krammer, K., Diatoms of Europe - Vol 4, *The genus Cymbopleura, Delicata, Navicymbula, Gomphocymbellopsis, Afrocymbella*, A. R. G. Gantner Verlag K. G., 2003.
7. Krammer, K. and Lange-Bertalot, H., Bacillariophyceae. Die Süsswasserflora von Mitteleuropa. Vol. 2/1 *Naviculaceae*, p.1-876 mit 206 pl. Vol. 2/2 *Bacillariaceae, Epithemiaceae, Surirellaceae*, p. 1-596 (1988). Vol 2/3 *Centrales, Fragilariaceae, Eunotiaceae*, p. 1-576 (1991). Vol. 2/4 *Achnanthaceae*, Kritische Erganzungen zu *Navicula* (Lineolatae) und *Gomphonema*. P. 1-437 (1991). Vol. 5 English and French translations of the keys and supplements. Stuttgart and Heidelberg.

8. Lange-Bertalot, H., Diatoms of Europe - Vol 2, *The genus Navicula sensu stricto* 10 Genera Separated from *Navicula* sensu lato *Frustulia*, A. R. G. Gantner Verlag K. G., 2001.

9. Metzeltin, D. Lange-Bertalot, H. and Garcia-Rodriguez, F., *Diatoms of Uruguay. Taxonomy-Biogeography-Diversity* (Annotated Diatom Micrographs ed Lange Bertalot H) Iconographia Diatomol 15, A. R. G. Gantner Verlag K. G., 2005.

10. Nautiyal, P.N. and Singh, H.R., Fish Diversity in the oriental highlands of the Indian subcontinent, ed. Biodiversity and Ecology of aquatic Environments, pg. 107-155, 2009.

11. Patrick, R. and Reimer, C.W., *The diatoms of the United States exclusive of Alaska and Hawaii*, I. Monographs of the Academy of Natural Sciences, Philadelphia, 1966.

12. Round, F.E., *The Ecology of Algae*. Cambridge University Press, New York, 1981.

13. Stevenson, R.J., Scale dependent determinants and consequences of benthic algal heterogeneity. J. N. Ameri. Benth. Soc., 1997; 16(24): 82-62.

14. Werum, M. and Lange-Bertalot, H., *Diatoms in springs from Central Europe and elsewhere* under the influence of hydrogeology and anthropogenic impacts. Ecology- Hydrogeology- Taxonomy, (H. Lange-Bertalot, ed.), 480. Iconographia Diatomologia, 13, A. R. G. Gantner Verlag K. G., 2004.

Floral Diversity and their Conservation (2013), Editors: D.R. Khanna et al.
Published by Biotech Books.
ISBN: 978-81-7622-286-0 Pages: 157-166

17

Soil and Moisture Conservation and Reafforestration Project

AMANDEEP KAUR GIRAN

The forests of Himachal Pradesh are rich in vascular flora, which forms the conspicuous vegetation cover. Out of total 45,000 species of plants found in the country as many as 3,295 species (7.32%) are reported in the State.

More than 95% of species are endemic to Himachal and characteristic of Western Himalayan flora, while about 5% (150 species) are exotic introduced over the last 150 years.

The forests of Himachal Pradesh are presently under great stress due to the impact of activities such as mining & construction, and growth in human and cattle populations.

The main objective of the project was to prevent soil erosion and increase the soil moisture content by means of construction of check dams, absorption pits, and afforestation. the area selected for this purpose was that of an abandoned mining site at Bhagsunag ,Dharamshala (Himachal Pradesh).

Local species of plants were planted so as to maintain the biodiversity of the area.

The result that have been achieved the past few years from 2007 are immense. the area was landslide prone but with the construction of check dams and plantation work the land has stabilized.

INTRODUCTION

The rich variety of life on Earth has always dealt with a changing climate yet climate change now poses as one of the principal threats to the biological diversity of the planet. The resilience of ecosystems can be enhanced and the risk of damage to human and natural ecosystems can be reduced through the adoption of biodiversity-based adaptive and mitigative strategies. Increasing the forest cover is one such strategy.

The relevance of forests is greater in a country like India where over a billion people are struggling to find breathing space. With nearly 200,000 villages classified as forest villages, there is obviously large dependence of communities on forest resources. India has 2.3 % of the world' s land area, 1.72% of the world's forests and supports 16.7% of the human population, 16% of world's goat and cattle population and 56% of world's buffalo population. As against a need of over 33% of forestland to maintain the ecological balance, the country today has only 21.02 % of forest-land. In terms of the quality of the forest cover, 2.54% is dense forest, 9.71% is medium dense forest, and 8.77% is open forest (SFR 2009). Also, as against a world average of 0.64 hectares, the per capita forest cover in India is only 0.064 hectares, thereby making the people of India living in a polluted atmosphere. Forest ecosystems in India are already subject to socio-economic pressures leading to forest degradation and loss, with adverse impacts on the livelihoods of the forest-dependent communities.

HIGH ALTITUDE FORESTS:

It is crucial to consider mountain regions as all of the world's rivers originate in them and flow to the oceans, sustaining the life of all beings. Though all forests play a crucial role in climate regulation, mountain forests are of paramount importance due to their regulation of the Earth's river sources. When these forests are extensively cut, massive problems relating to erosion occur and water tables diminish, leading to extreme drought. It is essential to look at what can be done to restore Earth's high altitude forests and to preserve what we have, while there is still time. Restoring mountain forests is particularly difficult when the old forests and top soils have been so severely diminished. Hence action should be prompt and intense. These forests need to be of indigenous biodiversity to function completely and effectively and therefore must be carefully managed, in order to maintain their ability to provide fresh water.

HIMACHAL PRADESH

The forests of Himachal Pradesh are rich in vascular flora, which forms the conspicuous vegetation cover. Out of total 45,000 species of plants found in the country as many as 3,295 species (7.32%) are reported in the State.

More than 95% of species are endemic to Himachal and characteristic of Western Himalayan flora, while about 5% (150 species) are exotic introduced over the last 150 years.

The Forests of Himachal Pradesh, known for their grandeur and majesty, are like a green pearl in the Himalayan crown. The forests of the State can be broadly classified, on an ecological basis as laid down by Champion and Seth, into Coniferous Forests and Broad-leaved Forests. Distribution of various species follows fairly regular altitudinal stratification. The vegetation varies from Dry Scrub Forests at lower altitudes to Alpine Pastures at higher altitudes. In between these two extremes, distinct vegetation zones of Mixed Deciduous Forests, Bamboo, Chil, Oaks, Deodar, Kail, Fir and Spruce, are found. The richness and diversity of the flora can be gauged from the fact that, out of the total 45,000 species found in the country, as many as 3,295 species (7.32%) are reported in the State.

These life supporting systems are presently under great stress. In Himachal Pradesh, nearly 1000 small to medium-sized slate mines have stripped upto 60 per cent of the forest and have triggered countless landslides.

KANGRA

Kangra district of Himachal Pradesh is situated in Western Himalayas between 31° 2 N to 32° 5 N and 75° E to 77° 45 E. The district has a geographical area of 5,739 km^2 which constitutes 10.31 per cent of the geographical area of the State. It has varying altitude ranging from 427m to 6401m above mean sea level. Kangra has considerable diversity in its soils, physiography. land use pattern and cropping system and has further been divided into five sub-situations *i.e.* Pir Panjal, Dhauladhar, Kangra Shiwalik, Kangra Valley and Bias Basin. The monsoon rains are heavy and well-distributed and 70 percent of the total annual rainfall is received from July - September.

The forest of district Kangra consists of vegetation right from Scrub forest at low elevation, to Alpine pastures at higher altitude. In-between interspersed are the forests of *chir*, *ban* oak, mixed conifers (kail, spruce and fur) and *kharsu* oak forests.

Table 1: Types of forests found in Kangra.

S. No	*Forest Type*	*Altitude (metres above sea level)*
1	Miscellaneous Scrub forest	600-1200
2	Chir forests	800-1700
3	Ban oak forests	1600-2300
4	Deodar forest	2000-2500
5	Mixed conifers	2100-3000
6	Kharsu oak forest	2300-3800
7	Alpine scrub and alpine pastures	Above 3800

DHARAMSHALA REGION

The town is situated on the southern slopes of Dhauladhar ranges. The northern slopes of Dharamshala have undergone rapid urbanisation and as a result have experienced extensive deforestation to accommodate residential, tourist and commercial establishments. These human interventions have laid bare sensitive slopes, which often fall during the monsoon and winter rains. Large tracts of land bear very thin soil cover that is unfit to support any type of vegetation. Hence there is an urgent need for scientific and systematic afforestation which will accelerate the process of re-establishment of forest in these areas.

OBJECTIVES

- To formulate a method that will help in increasing the vegetation cover of selected degraded areas at high altitude, by using the most suited techniques.
- To prevent soil erosion and increase the soil moisture content by means of construction of check dams, absorption pits, gabion structures etc., and afforestation.

SELECTION OF AREA:

Three sites were recommended by the officials of Forest Department for the project. Achieving results on these sites posed as a real challenge as these sites were highly degraded and difficult to work on and hence were not taken up by the Forest Department.

Site One:

This site, situated at Gallu, is on the route that is taken twice a year by the migratory grazers. The approach to the said area, especially the last 3 kms., is extremely difficult.

Site Two:

This site, at Dharamkot, admeasuring around 4 hectares has been subjected to lime-stone quarrying for a prolonged period.

Site Three:

This site is situated near Bhagsunag Temple at Bhagsunag admeasuring 270 hectares and has been severely degraded due to slate mining.

This site, being close to a waterfall that is visited by large number of people during tourist season, would help in creating awareness. This site, being close to the village, is regularly visited by the forest guards, and is better protected. Hence this area was selected for the soil water conservation and afforestation work.

WORK PROPOSED TO BE CARRIED OUT

The work of soil moisture conservation and afforestation of the selected area will be carried out in several phases. Each phase will comprise of Soil Moisture Conservation (SMC) and Afforestation.

A. Soil Moisture Conservation

Water is essential for life and is also part of the larger ecosystem on whic biodiversity depends. Precipitation, converted to soil water and groundwate and thus accessible to vegetation and people, is the dominant pre-conditio for biomass production and social development.

Techniques Used For SMC

A variety of essential soil moisture and water conservation techniques mus be adopted to increase the moisture content of the soil. Techniques can b selected and modified based on the conditions prevailing on the respectiv sites.

1. **Check dams:** Check dams are the most suited technique for SMC in areas that are very steep and have very shallow soil cover. These can be temporary or permanent structures.
2. **Percolation Ponds:** A percolation pond is constructed by excavating a depression, forming a small reservoir, or by constructing ar embankment in a natural ravine or gully to form an impoundec type of reservoir.
3. **Contour bunds:** This is a suitable method to reduce the run-of loss of water and increase the effectiveness of the rainfall in areas where the elevation is more and most of the rainfall is lost through run-off.
4. **Absorption Pits:** Depending on the gradient of the land, pits are dug at regular intervals to allow collection of rain-water that is slowly absorbed by the soil thereby increasing the soil moisture.

Depending on the topography of the area and the conditions prevalent, the techniques mentioned above can be used singly or in combination for effective Soil Moisture Conservation.

SMC At Project Site

After a detailed study of the project area, check dams were selected as the most suitable technique for Soil Moisture Conservation (SMC). Check dams built by compactly piling stones are called *Dangi* and are long-lasting. Stones lying around the area were used for the construction. It will serve 3 nain purposes. Firstly it will impart stability to the entire stretch of the land; econdly it will prevent goats from entering the site, and thirdly, it negates he requirement of fencing that large stretch.

It is also cost-effective and does not cause an undue increase in the veight of the hill. It also helps in prevention of soil erosion & loss & etention of water.

STEPS FOLLOWED

A. Dangi Construction:

The existing natural contours were demarcated for *Dangi* construction and cleared to create a base of 2 feet to 2.6 feet for the *Dangis*. Care was aken to see that the construction had minimal negative effect on the natural opography.

Locals of Bhagsunag, who are experienced in constructing *Dangis* vere employed for the said work. Depending upon the gradient and soil, *Dangis* of varying heights (2 feet to 8 feet) were constructed. A concrete ayer of 2 to 3 inches was laid down on every *Dangi* to add further stability o the said structure. After the *Dangi* construction was completed, the slate and rubble lying in the area was cleared and added at the base of the *Dangis*, which further helped in strengthening it.

B. Afforestation

1. Fencing

The area was fenced using 40mm x 40mm x 6mm L iron angles and with ive strands of barbed wires to prevent the entry of grazing animals in the area. Further fortification of the fencing was done by interlacing the barbed vires with branches of Barberry.

2. Plantation

Plantation was carried out at a rate of 1100 trees per hectare, as this is the rate followed by the Forest Department. Good quality cow-dung manure was added to the pits of dimension 1x1x1.5 feet (prepared for plantation). Some pits were left unplanted so that they would serve as absorption pits.

Indigenous plant species were used so that the ecological balance of he area and the biodiversity of the forest remain undisturbed as much as possible. Plantation was carried out with the onset of monsoon.

Table 2: Some of the species used for plantation are listed below:

S.No	*Botanical Name*	*Common Name*	*2007*	*2008*	*2009*	*2011*
1	*Prunus pardus*	Pajja	-	40	100	**100**
2	*Quercus leucotrichophora*	Ban	500	50	200	**250**
3	*Aesculus hippocastanum*	Akhnor	-	30	-	-
4	*Cedrus deodara*	Deodar	10	50	350	**350**
5	*Albizia chinensis*	Ohi	-	30	50	-
6	*Poplar* spp	Poplar	-	20	45	-
7	*Soymida* spp	Dhuri	-	30	80	**60**
8	*Juglans* spp	Walnut	25	-	-	-
9	*Pinus roxburghii*	Chil	25	-	-	-
10	*Salix* spp	Willow	300	-	-	-
11	*Robinia pseudoaccacia*	Robinia	150	-	-	-
12	*Grevillea robusta*	Silver oak	-	-	-	**50**

The area was sprayed with seeds of the following species of grasses as they are effective soil binders and provide additional vegetation cover to the area.

Table 3: Grass Seeds Recommended by Palampur Agricultural University

S.No	*Botanical Name*	*Common Name*
1	*Festuca arundinacea*	Tall Fescue
2	*Dactylis glomerata*	Orchard Grass
3	*Trifolium repens*	White Clover

The area enclosed this year is approximately one acre and the area enclosed so far admeasures approximately 4 acres. A single main *Dangi* of dimension 180 feet x 2 feet x 4 feet was constructed at the base of the site along with seven small *Dangi's* admeasuring 160 feet in length in the path of heavy run-off during the monsoon. Fencing was done on the southern side of the site. Plantation work was started on 30th June, 2011 and was completed on 3rd July, 2011. Total number of plants planted in this phase was 810.Cow-dung manure was added at the base of the plant. Deodars, Pajja, Ban and Dhuri were planted at the site. A new species, Silver oak was also planted this year as suggested by the Forest Department.

RESULTS OBTAINED FROM THE PROJECT

Prior to the initiation of the project, the area experienced frequent landslides and heavy run-off during the monsoon. Soil erosion not only caused the rocks to move, causing heavy damage to the sparse vegetation, but also caused damage to the forest present on the upper periphery of the site. Post completion of two phases, substantial numbers of *Rhododendrons*, along with shrubs like *Rubus, Berberis lycium, Spiraea canescence* etc. have established themselves naturally on the site indicating that the water retention capacity as well as stability of the site has increased. Plants like *Rumex, Trifolium repens* etc. have also established themselves. 75% survival rate has been achieved. The forest on the upper periphery of the site has started moving downwards and hence natural regeneration is being provided a boost by this project. Determination of the results of plantation of 2011 will be possible only after October 2011.

BENEFITS ARISING FROM THE PROJECT

Direct Benefits:

- Has and will generate employment for a number of locals during the project
- Has and will provide permanent employment to one local, who has been appointed as the care taker of the project site
- The area can be opened to the locals to meet their need of fuel and fodder, once the trees are established

Indirect Benefits:

- Ground stability as the occurrence and intensity of land slides has reduced
- Substantial increase in the vegetation cover of the entire hill indicating that the land has stabilised and water retention capacity of the soil has increased substantially
- Increase in the biodiversity of the area
- Increase in the aesthetic value of the area and hence more inflow of tourists to the area

ACKNOWLEDEGEMNTS

I wish to thank CAT (Conservation Action Trust) for the grant for this project. I also wish to extent my gratitude to the Himachal Pradesh Forest Department officials. This project would not have been completed successfully without the help of local people of Bhagsunag. Last but not the least I wish to thank my family and friends for their continues support .

REFERENCES

1. Dadhwal VK(2009). Monitoring Forest for sustainability - Remote Sensing studies in India. *CAETS 2009*. Calgary
2. http://hpforest.nic.in/frst2.htm
3. Ravindranath NH, Joshi NV, Sukumar R and Saxena (2006). An Impact of climate change on forests in India. *Current Science*, 90:3
4. Ravindranath NH and Sudha P (2004). Joint Forest Management in India: Spread, Performance and Impacts, Universities Press, Hyderabad.
5. Rawat YS, Jina BS, Sah P and Bhatt MD (2008). Estimating Carbon Sequestration Rates and Total Carbon Stockpile in Degraded and Non-Degraded Sites of Oak and Pine Forest of Kumaun Central Himalaya. *ECOPRINT* 15: 75-81
6. State of Forest Report (2009). *Forest Survey of India*, (Ministry of Environment and Forest), Dehradun.

Floral Diversity and their Conservation (2013), Editors: D.R. Khanna et al.
Published by Biotech Books.
ISBN: 978-81-7622-286-0 Pages: 167-177

18

Effect of Eco-friendly Post Harvest Operations on Seed Quality of Green Gram (*Vigna radiate L*)

Sanjeev Charjan[1], Rajesh Gadewar[2], Ashish Lambat[2]

[1]Assistant Professor, Dr. P D K V's, College of Agriculture, Nagpur, M.S, India

[2]Assistant Professor, Sevadal Mahila Mahavidyalaya, Nagpur, M.S, India

Effect of eco-friendly threshing, drying and storage methods on seed quality of greengram were studied. The seeds threshed by hand had maintained significantly very low mechanical damage to seeds and higher germination, seedling vigour and field emergence percentage. Drying of greengram seeds at lower temperature (Shade) recorded significantly higher germination, seedling vigour and field emergence percentage. The greengram seeds stored in polyethylene bags recorded significantly higher germination, seedling vigour, field emergence and lesser seed invasion by fungal flora during storage as compared to jute and cloth bags under ambient condition. It was also noted that *Alternaria* sp., *Aspergillus* sp., *Fusarium* sp., *Rhizopus* sp. *Tricothecium* sp. and *Penicillium* sp. were the most commonly occurring fungi irrespective of storage periods and storage containers during storage under ambient condition.

Key words: Eco-friendly, Germination percentage, seedling vigour, field emergence percentage, storage, greengram.

INTRODUCTION

Greengram (Vigna radiata L.) is a rich source of protein but it is yet to gain impetus in our country. One of the major problems encountered in *Greengram (Vigna radiata L.)* production in India is the lack of good quality seeds. Seed yield and its quality depend on a number of factors. The time of harvesting, methods of threshing, drying and storage of seeds are among the major consideration in deciding the seed quality and productivity. Seed quality may be impaired while the seeds are still on mother plant (Pullock, 1972). Physio-morphological and physiological changes might set in, if the seeds are retained on the mother plant for a longer duration beyond physiological maturity (Ovchrov and Kizilova, 1966). Dharmalingam and Basu (1988) reported that the seed harvested at physiological maturity had highest germination and vigour than those harvested at premature. Sangakkara and Wanisekera (1990) reported that the mungbean seed quality not affected by hand threshing, drying at low temperature and storing the seeds in polyethylene bags in low temperature and relative humidity. As very little information is available on this aspect, hence an experiment was planned to study the effect of post-harvest operations on seed quality of *Greengram (Vigna radiata L.)*.

MATERIALS AND METHODS

Plants of *Greengram (Vigna radiata L. Cv. S-8)* were harvested at physiological maturity (*i.e.* when green pod turns in blackish green in colour) by cutting to a height of 5 cm from ground level. Pods were removed from the harvested plants and dried on the threshing floor as per usual method of drying. The harvested pods were divided in to 5 lots. First lot was threshed by hand, 2nd by stick beating 3rd by machine threshed 300 rpm. 4th by 400 rpm and 5th by 500 rpm separately. Half threshed seeds of each threshing methods dried in the sun and remaining half threshed seed was dried in shade. Three days were required for sun drying and 10 days were required for shade drying to reach safe moisture content level (10 $\pm$ 1 %). The threshed and dried seeds by different methods were kept in jute, cloth and polyethylene bags of 178 μm of size 20cm x 30cm respectively of 1Kg each and stored in wire – mesh almirah in masonry building having cemented wall, roof and floor under ambient condition for 9 months. The observations on moisture content and germination tests was conducted as prescribed in International rules for seed testing (ISTA, 1985). Two hundred seeds were used to isolate fungi following standard blotter and agar plate methods (ISTA, 1976). The seedling vigour index was worked out following the method of Aldul Baki & Anderson (1973). For filed emergence test, sowing of *Greengram (Vigna radiata L.)* seeds was done in randomized block design, with four replications with inter and intra-row spacing of 1 feet and 6 inches respectively. Observations for field emergence were recorded daily and finally the established seedlings were counted after one month of sowing. The experimental data was statistically scrutinized by random block design as per Panse and Sukhatme (1967).

RESULTS AND DISCUSSION.

Data presented in the Table 1 showed that machine threshed at 500 rpm results into seeds with significantly maximum mechanical damage (3.4%) which is followed by 400 rpm (2.7%), 300 rpm (1.8%), stick beating (1.3%), and hand threshed (0.3%). Saini *et al.*. (1982), Sangakkara and Wanisekera (1990) reported that the seed threshed by hand lower mechanical damage to seed coat than machine threshed. The *Greengram (Vigna radiata L.)* crop threshed by hand showed significantly highest sound seed (99.8%), germination (91%), Seedling Vigour Index (1026) and Field emergence (83%) as compaired to other threshing methods.

In was observed from the Table 2 that the germination percentage of *Greengram (Vigna radiata L.)* seeds threshed by different methods varies significantly and it was highest in hand – threshed seeds (90 %) which was closely followed by stick- beating (87%), machine threshed 300 rpm (85%), 400 rpm(82%) and 500 rpm (77%). The germination percentage was decreased with increasing in mechanical damage to the seed coat of *Greengram (Vigna radiata L.)*. It was also observed that the abnormal seedling increases with the increasing the mechanical damage to the seed coat of *Greengram (Vigna radiata L.)*. The similar results also reported by Soesarsano and Copeland (1974), Sangakkara and Wanisekera (1990). Seedling vigour index and filed emergence followed the same trends of germination. It was highest in hand- threshed seeds as compared to other threshing method.

Table1: Effect of threshing methods on mechanical damage to seed coat, sound seeds and germination percentage in *Greengram* (*Vigna radiata L.*)

S. No.	*Threshing methods*	*Mechanically damaged seed (%)*	*Sound seed (%)*	*Germination analysis %*			*Seedling Vigour index*	*Field emergence (%)*
				Normal seedling (%)	*Abnormal Seedling (%)*	*Dead seed (%)*		
1	Hand - threshing	0.3	99.6	93	2	5	1046	85
2	Stick Beating	1.3	98.7	89	6	5	1000	80
3	Machine - threshing 300 rpm	1.8	98.2	81	9	10	919	70
4	400 rpm	2.7	97.7	78	14	8	890	66
5	500 rpm	3.4	96.6	74	14	12	820	60
	SE (m) ±	0.3	0.3	1.3	n/a	n/a	70	1.4
	C.D. at 5 %	1	1	4.1	n/a	n/a	218	4.3

Germination % on the basis of normal seeding % (ISTA.,1985). SE (m)± Standard Error mean, C.D. at 5% Critical Difference at 5% in Random Block Design.

Table 2: Effect of threshing and drying methods on germination percentage, seedling vigour index and field emergence percentage in *Greengram* (*Vigna radiata L.*).

S. No.	*Threshing Methods*	*Shade drying*					*Sun drying*				
		Germination Analysis %					*Germination Analysis %*				
		NS	*ABS*	*DS*	*SVI*	*FE %*	*NS*	*ABS*	*DS*	*SVI*	*FE %*
1	Hand- threshing	90	6	4	1076	83	86	7	7	902	77
2	Stick - beating	86	9	5	987	76	80	11	9	873	70
3	Machine - threshing 300 rpm	85	10	5	962	72	75	15	10	786	64
4	400 rpm	82	12	6	894	68	70	18	12	755	57
5	500 rpm	77	17	6	804	62	62	24	14	654	48
	SE (m) ±	1.4	n/a	n/a	55	1.2	1.2	n/a	n/a	60	1.5
	C.D. at 5 %	4.2	n/a	n/a	168	3.6	3.7	n/a	n/a	180	4.5

(G.A- Germination analysis, NS- Normal Seedling, ABS- Abnormal seedling, DS- Dead seed, SVI-Seedling vigour index, FE-field emergence. SE (m)± Standard Error mean, C.D. at 5% Critical Difference at 5%.)

It was observed from Table 3 that the germination of sundried hand threshed seeds was (85%) which is closely followed by stick beating (80%), machine threshed 300 rpm (75%), 400 rpm (70%) and 500 rpm (62%). As against 90% in hand threshed, 86 % in stick beating, 85 % in machine threshed 300 rpm 82% in 400 rpm and 77% in 500 rpm in shade dried seeds. In shade drying, low temperature (23-30°C) minimizes scorching due to gradual loss of the moisture and there was no adverse effect on germinability of seeds. The result confirms the findings of Philpot, (1976), Patil and Zode, (1993). In sun drying, high ambient (38-42°C) temperature develop a rapid flow of moisture within the seeds causing stress and sun scorching to the embryo and cotyledons and this results in reduction of germinability with high number of abnormal seedlings and dead seeds Morrison and Robertson, (1978), Sangakkara and Wanisekera, (1990), Patil and Zode, (1993). Seedling vigour index and field emergence percentage also followed the same trend of germination in *Greengram (Vigna radiata L.)*. Similar results also reported by Saini *et al.*. (1982), Patil and Zode, (1993).

Table 3: Effect of threshing, drying methods and storage containers on germination percentage, seedling vigour index, field emergence percentage in Greengram (*Vigna radiata L.*) during storage.

S. No.	*Drying Methods / Packing Methods*	*Germination %*		*Seedling vigour index*		*Field Emergence %*	
		Initial	*9 Months*	*Initial*	*9 Months*	*Initial*	*9 Months*
1	**Hand threshed**						
	I) Shade drying						
	i) Jute bag	90	83	1076	909	83	69
	ii) Cloth bag	92	84	1076	920	83	71
	iii) Polyethylene bag	92	91	1076	1062	83	81
	II) Sun drying						
	i) Jute bag	85	69	902	752	76	56
	ii) Cloth bag	85	70	902	769	76	58
	iii) Polyethylene bag	85	82	902	890	76	72
	SE (m) ±	**n/a**	**24**	**n/a**	**41**	**n/a**	**2.2**
2	**Stick beating**						
	I) Shade drying						
	i) Jute bag	86	79	937	861	76	66

contd...

S. No.	Drying Methods / Packing Methods	Germination %		Seedling vigour index		Seedling vigour index	
		Initial	9 Months	Initial	9 Months	Initial	9 Months
	ii) Cloth bag	87	80	937	874	76	68
	iii) Polyethylene bag	87	85	937	906	76	75
	II) Sun drying						
	i) Jute bag	80	70	873	772	70	56
	ii) Cloth bag	80	72	873	794	70	60
	iii) Polyethylene bag	80	76	873	852	70	66
	SE (m) ±	**n/a**	**3**	**n/a**	**42**	**n/a**	**2.8**
3	**Machine- threshed (300 rpm)**						
	I) Shade drying						
	i) Jute bag	85	70	912	764	72	56
	ii) Cloth bag	85	71	912	771	72	57
	iii) Polyethylene bag	85	80	912	882	72	68
	II) Sun drying						
	i) Jute bag	75	58	840	8650	64	46
	ii) Cloth bag	75	60	840	662	64	48
	iii) Polyethylene bag	75	70	840	769	64	60
	SE (m) ±	**n/a**	**3.1**	**n/a**	**44**	**n/a**	**3**
4	**Machine- threshed (400 rpm)**						
	I) Shade drying						
	i) Jute bag	82	62	894	662	68	50
	ii) Cloth bag	82	64	894	686	68	51
	iii) Polyethylene bag	82	68	894	738	68	55
	II) Sun drying						
	i) Jute bag	70	56	761	542	57	42
	ii) Cloth bag	70	58	761	607	57	43
	iii) Polyethylene bag	70	64	761	720	57	54
	SE (m) ±	**n/a**	**2.8**	**n/a**	**40**	**n/a**	**2.6**
5	**Machine- threshed (500 rpm)**						
	I) Shade drying						
	i) Jute bag	77	55	867	534	62	40
	ii) Cloth bag	77	56	867	542	62	41
	iii) Polyethylene bag	77	67	867	752	62	51
	II) Sun drying						
	i) Jute bag	62	40	654	362	48	25
	ii) Cloth bag	62	45	654	400	48	30
	iii) Polyethylene bag	62	50	654	468	48	36
	SE (m) ±	**n/a**	**3.7**	**n/a**	**59**	**n/a**	**4**

SE (m) ± Standard Error mean,

Table 4: Effect of threshing, drying methods and storage containers on incidence percentage of fungal flora on *Greengram* (*Vigna radiata L.*) during storage.

S. No.	*Treatments*	*Percentage of fungi encountered on Greengram (Vigna radiata L.) seeds*											
		A		B		C		D		E		F	
		1	2	1	2	1	2	1	2	1	2	1	2
1	**Hand threshed**												
	I) Shade drying												
	i) Jute bag	1	6	3	14	1	4	1	20	n/a	6	n/a	1
	ii) Cloth bag	1	6	3	10	1	6	1	22	n/a	5	n/a	n/a
	iii) Polyethylene bag	1	2	3	6	1	1	1	4	n/a	3	n/a	n/a
	II) Sun drying												
	i) Jute bag	1	5	2	15	n/a	7	n/a	10	n/a	5	n/a	1
	ii) Cloth bag	1	5	2	16	n/a	8	n/a	16	n/a	5	n/a	1
	iii) Polyethylene bag	1	3	2	7	n/a	1	n/a	12	n/a	2	n/a	n/a
2	**Stick beating**												
	I) Shade drying												
	i) Jute bag	2	6	2	15	n/a	8	1	24	1	7	n/a	2
	ii) Cloth bag	2	7	2	16	n/a	9	1	20	1	9	n/a	1
	iii) Polyethylene bag	2	1	2	10	n/a	4	1	10	1	2	n/a	n/a
	II) Sun drying												
	i) Jute bag	1	4	2	20	n/a	10	n/a	20	n/a	3	n/a	1
	ii) Cloth bag	1	4	2	14	n/a	9	n/a	24	n/a	4	n/a	2
	iii) Polyethylene bag	n/a	n/a	2	8	n/a	2	n/a	14	n/a	n/a	n/a	n/a
3	**Machine- threshed (300 rpm)**												
	I) Shade drying												
	i) Jute bag	1	5	3	18	1	8	1	27	1	9	n/a	n/a
	ii) Cloth bag	1	6	3	16	1	5	1	24	1	8	n/a	n/a
	iii) Polyethylene bag	1	1	3	9	1	1	1	12	1	2	n/a	n/a
	II) Sun drying												
	i) Jute bag	1	5	3	20	n/a	6	1	24	n/a	18	n/a	n/a
	ii) Cloth bag	1	5	3	18	n/a	4	1	14	n/a	15	n/a	n/a

contd...

	iii) Polyethylene bag	n/a	1	3	10	n/a	n/a	1	8	n/a	2	n/a	n/a

S. No.	***Treatments***	***Percentage of fungi encountered on Greengram (Vigna radiata L.) seeds***											
		A		**B**		**C**		**D**		**E**		**F**	
		1	2	1	2	1	2	1	2	1	2	1	2
4	**Machine- threshed (400 rpm)**												
	I) Shade drying												
	i) Jute bag	1	6	3	27	1	14	1	24	1	9	n/a	1
	ii) Cloth bag	1	6	3	19	1	9	1	19	1	8	n/a	2
	iii) Polyethylene bag	1	1	3	16	1	1	1	2	1	2	n/a	n/a
	II) Sun drying												
	i) Jute bag	1	6	2	31	n/a	19	n/a	25	n/a	10	n/a	2
	ii) Cloth bag	1	5	2	29	n/a	14	n/a	19	n/a	10	n/a	2
	iii) Polyethylene bag	0	n/a	2	18	n/a	5	n/a	3	n/a	3	n/a	n/a
5	**Machine- threshed (500 rpm)**												
	I) Shade drying												
	i) Jute bag	1	6	3	27	1	19	1	30	1	15	n/a	n/a
	ii) Cloth bag	1	5	3	21	1	12	1	25	1	10	n/a	1
	iii) Polyethylene bag	1	2	3	12	1	12	1	15	1	2	n/a	n/a
	II) Sun drying												
	i) Jute bag	1	5	2	28	1	19	1	34	n/a	15	n/a	1
	ii) Cloth bag	1	4	2	20	1	12	1	29	n/a	9	n/a	2
	iii) Polyethylene bag	n/a	n/a	2	10	1	7	1	11	n/a	2	n/a	n/a
		29	106	65	318	15	150	21	424	12	155	0	20

A- *Alternaria* **sp., B-***Aspergillus* **sp. C-** *Fusarium* **sp., D-** *Penicillium* **sp., E-** *Rhizopus* **sp. and F-** *Trichothecium* **sp.,**

1- Standard blotter paper method test, 2- Agar plate method.

The apparent influence of threshing, drying methods and storage containers on germination, seedling vigour index and field emergence percentage of *Greengram (Vigna radiata L.)* seeds during storage presented in Table 4. The data indicated that the germination, seedling vigour and field emergence percentage of stored *Greengram (Vigna radiata L.)* declined with increasing storage period, however the rate of loss varied with the methods of threshing, drying and types of storage containers used. A sharp declined in germination, seedling vigour index and field emergence occurred in seeds threshed by machine 500 rpm, drying in sun and stored in jute, cloth and polyethylene bags during storage. However, there was practically less loss of germination, seedling vigour index and field emergence percentage in seeds threshed by hand dried in shade and stored in jute, cloth and polyethylene bags during storage for 9 months Sangakkara and Wanisekera, (1990), Patil and Zode, (1993) also reported that the sun drying of seeds harmful for germination. In the present study, it was observed that the *Greengram (Vigna radiata L.)* seed, stored in polyethylene bags were undergo the least amount of loss of germinability to great extent as compared to jute and cloth bags irrespective to threshing and drying method Vanangamudi, (1988), Shivankar *et al..*, (1990), Likhitkar and Charjan, (1995) also reported superiority of polyethylene bags over jute and cloth bags for successful carry over of seeds during storage. In the present study, the similar trends of germination were observed in seedling vigour index and field emergence percentage of Greengram *(Vigna radiata L.)* seeds during storage.

During this study (Table 4) it was evident that the maximum number of fungal colonies did develops in jute bags which are followed by cloth bags and polyethylene bags during storage. The polyethylene bags provided much protection as polyethylene bags resist moisture penetration which helps in preventing the development of fungal colonies both quantitative and species wise irrespective to threshing and drying methods. It was also noted that *Alternaria* sp., *Aspergillus* sp. *Fusarium* sp., *Penicillium* sp., *Rhizopus* sp. and *Trichothecium* sp. were the most commonly occurring fungi irrespective of storage periods and containers. The germinability of *Greengram (Vigna radiata L.)* seeds decreased with increase in incidence percentage of fungal flora and storage period. The isolated fungi were most inhibitory to germination,. The results obtained were in conformity with the findings of Charjan and Gupta (1996) & Wankhede et.al (2010). It was also observed that the lower storage potential of mechanical damage seeds threshed by machine and sun dried may be because of higher rate of respiration and attraction of more fungal flora due to higher leaching of sugar which cause early reduction in germination, seedling vigour and field emergence (Burriga, 1961).

CONCLUSION.

Thus, these results highlight the maintenance of *Greengram (Vigna radiata L.)* seed quality for sowing purposes in the next sowing reason for getting better yield. The *Greengram (Vigna radiata L.)* pods should be threshed by hand or stick beating and dried in shade upto safe moisture level (10 $\pm$ 1%) and stored in polyethylene bags, shows greater germinability, seedling vigour index and filed emergence and lesser invasion by fungal flora upto the next sowing season.

REFERENCES

1. Abdul Baki A. and J. D. Anderson. 1973. Vigour determination in soybean seed by multiple criteria. Crop Sci. 13 : 630 – 633.
2. Barriga, C. 1961. Effect of mechanical abuse of navybean seed at various moisture levels. Agron J. 63 : 250 – 250
3. Charjan, S. K. U. and V. R. Gupta. 1996. Impact of storage condition on fungal flora and germinability of gram seeds. J. Soils and crops 6 (2) : 136 – 138
4. Dharmalingam, C. and R. N. Basu. 1988. Seed quality in relation to position of seed in the pod at different maturity periods in mungbean. Seed Research. 16 (2) : 168 – 172
5. ISTA. 1976. International rules for seed testing seed sci. an technol 4 : 108
6. ISTA. 1985. International rules for seed testing seed sci. and Technol 13: 299-513
7. ISTA, 1993. International rules for seed testing, Seed Sci. & Technol 21; 25-30.
8. Likhitkar, V. S. and S. K. U. Charjan. 1995. Impact of storage containers and period on germination of Gmelina arborea seeds during storage. In : Herval medicines, biodiversity and conservation strategies (Ed. R. C. Rajak and M. K. Rai), international book distributors, Distributors, Deharadun : 236 – 241.
9. Morrison, W. H. III and J. A. Robertson. 1978. Effect of drying on sunflower seed oil quality and germination. J. Am. Oil Chem. Soc. 55 : 272-274.
10. Ovcharov, K. E. and K. G. Kizilova 1966. Difference in seed quality and plant productivity. In : Physiological basis of seed germination. Amerinol publishing, New Delhi pp. 140.
11. Panse, V. G. and P. V. Sukhatme. 1967. Statistical methods for agril. Workers I. C. A. R. Publication New Delhi.

12. Patil, V. N. and N. G. Zode. 1993. Effect of seasons and drying methods on storability of ground nut (arachis hypogaea) Seed Research. Special Vol. 1: 342 – 347.

13. Philpot, R. 1976. Principles and practices of seed drying. Proceeding of Mississipi state University Short cource for seedmen 16 : 23 – 40.

14. Pullock, B. M. 1972. Effects of environment after sowing on viability. In: Viability of seeds (Ed. E. H. Roberts), Chapman and hall ltd. London pp. 150-171.

15. Saini, S. K. , J. N. Singh and P. C. Gupta. 1982. Effect of threshing method on seed quality of soybean. Seed Research. 10 (2) : 133 – 138.

16. Sangakkara, U. R. and w. M. T. Wanisekera. 1990. Effect of post- harvest operations on seed quality of mungvean. Seed Research. 18 (1) : 54-59.

17. Shivankar, V. J., S. N. Singh, A. A. Khan and A. S. Tomer. 1990. Effect of storage conditions on storability of cowpea. In International conference on seed science and technology, New Delhi.

18. Soesarsano, W. and I. C. Copeland. 1974. Effect of original moisture content, maturity and mechanical damage on seed and seedling vigour of beans. Agron J. 66 : 546 – 548.

19. Vanangammudi, K. 1988. Storability of soybean seeds as influenced by the variety, seed size and storage container. Seed Research. 16 (1) : 81-87.

Floral Diversity and their Conservation (2013), Editors: D.R. Khanna et al.
Published by Biotech Books.
ISBN: 978-81-7622-286-0 Pages: 179-190

19

Study of Phytoplankton Density of Sita pat Pond, Dhar

CHOUDHARY PREET[1]., WASKEL D.S.,[2] BAGHEL LAXM[3] AND RAWAT RANJANA[4]
[1, 2, 3]Department of zoology, Govt. P.G. College, Dhar (M.P.)
[4]Department of zoology, Govt. P.G. College, Zhabua (M.P.)

The plankton plays a very important role for maintaining the productivity of the water body. Plankton refers to microscopic aquatic plants having little or no resistance to water current and living free floating and suspended in open or "pelagic waters". Phytoplankton's 24 species belonging to four groups Chlorophyceae, Cyanophyceae, Bacillariophyceae and Euglenophyceae were identified from sitapat pond, Dhar (M.P.). These groups are represented in order of dominance as Chlorophyceae>Cynophyceae> Bacillariophyceae > Eugleuophyceae.

Key words: Phytoplankton's, Productivity, Pelagic-waters.

INTRODUCTION

Plankton refers to microscopic aquatic plants or animals having little or no resistance to water current and living free floating and suspended in open or "pelagic waters". Plank tonic plants are called as phytonectons. They play a significant role in aquatic system as consumers.

Phytoplanktons have, since long, served useful in monitoring water quality in aquatic systems. Studies on polluted systems with reference to various pollutants have thrown light on the effectiveness plankton as bio-indicators (Jaseph & Pillai 1975, Tripathi 1985, Raman & Phani Prakash 1989, Gupta & Sukhla1990). The plankton population was identified up to genus level and regrouped in to the various groups as seasonal variations of total phytoplankton in the Sitaput pond of Dhar town (m.p.) was studied in the year of 2007-2008.

MATERIAL AND METHODS

The plankton samples were collected through fine plankton net. Water samples filtering 40 liters of water through small plankton net made up of bolting silk no. 25 (64u mesh size). The concentrate was preserved in 5% formalin and Lugol's solution for phytoplankton and zooplankton study respectively. The phytoplankton was identified with the help of key's given by Smith (1950), Edmondon (1959), Prescott (1951 and 62) and Adoni (1985). Counting of the individual phytoplankton was done by drop count method (Adoni, 1985) using the formula

Phytoplankton / lit. = A x 1/L x n/V

Where; A = Average no. of organism /drop

L = Volume of original sample in ml

n = Volume of one drop in ml.

V = Total volume of the concentrated sample in ml

$$\text{Mean } (\bar{X}) = \frac{\sum X}{N}$$

where, $(\bar{X})$ = Individual reading of parameter

$\sum X$ = Total of all readings for parameter

N = Number of samples

Standard Deviation (SD):

$$SD = \frac{\sqrt{\sum (X-\bar{X})^2}}{n-1}$$

Density = mean + SD×10^2

Where, X = Individual reading of parameter

$\bar{X}$ = Mean of $\sum X$

n = Number of samples

RESULTS

In the present study, the phytoplankton population was found to be comprising of four major groups *viz.* Clorophyceae, Cyanophyceae, Bacillariophyceac and Englenophyceae. In all 24 species of phytoplankton were identified out of which 11 to Chlorophyceae, 6 to Cyanophyceae, 5 to Bacillariophyceae and 2 to Euglenophyceae respectively.

In the present study the density of total phytoplankton 4268 No/lit and 4103 No/lit. During 2007 and 2008 respectively. In the Sitapat pond the succession of phytoplankton is noticed as Chlorophyceae>Cyanophyceae> Bacillariophyceae > Englenophyceae.

Chlorophyceae:

In the present study in 2007 and 2008, chlorophyceae group (green algae) dominated the total phytoplankton forming in Sitapat pond. The chlorophyceae dominated in the number of genera and species; however their percentage composition was ranged between 11.87% to 20.62%. The peak of the group was recorded in the summer season and minimum in the rainy season during the years of investigation.

Cyanophyceae:

The cyanophyceae (blue green algae) is an important part of phytoplanktons in Sitapat pond .In the present study in 2007 and 2008, Cyanophyceae comprised second dominant group and ranged between 6.87% to 9.64%. The maximum density observed during summer season while minimum density observed during rainy season.

Bacillariophyceae:

The bacillariophyceae constitituted important part of phytoplankton in Sitapat pond. In the present study in 2007 and 2008, group bacillariophyceae constituted 6.42% to 8.38% of the total phytoplankton. The minimum density of bacillariophyceae was observed during the summer season while the maximum density of the same was observed during the winter season.

Euglenophyceae:

In the present study the percentage composition of group Euglenophyceae was 1.11% to 2.31%. This group is represented by two species Euglena and Trachelomonas. The maximum density observed during the rainy season while minimum density was observed in the summer season in the study period.

DISCUSSION

Chlorophyceae:

The chlorophyceae comprise of a very large number of diverse forms which enjoy a wide distribution in aquatic and terrestrial habitats. Some of the fresh water planktons are found on the surface of stagnant water or grow firmly attached to submerged rocks, plants and similar objects in free flowing streams, but some are marine, while others prefer moist terrestrial habitats, such as dump soil and the shaded sides of rocks and bark of trees.

In the present study in 2007 and 2008, chlorophyceae group (green algae) dominated the total phytoplankton forming in Sitapat pond. The chlorophyceae dominated in the number of genera and species; however their percentage composition was 11.87% in 2007 and 20.62% in 2008. The peak of the group was recorded in the summer season and minimum in the rainy season.

In the present investigation, it was observed that high temperature and pH were favorable for rapid development of chlorophyceae. The chlorophyceae are the indicators of healthiness of the ecosystem and are also responsible for addition of high amounts of oxygen to the water body. Similar observations were also made by Tripathi and Pandey (1990), Brajesh and Pandey (2001), Sunitha *et al.*. (2005), Susheela and Kiran (2006).

Cyanophyceae:

The cynophyceae are known as a successful group and enjoy a wide distribution in all kinds of habitats. They are perhaps of greater ecological importance as pioneer forms than the members of any other class of algae.

The cyanophyceae is (blue green algae) an important part of phytoplanktons. Cyanophyceae comprised second dominant group and ranged between 6.87% to 9.64%. The maximum density was observed during summer season while minimum density was observed during rainy season, when Phosphate and BOD were higher in these seasons. Similar observations were also made by Sreenivasan (1964), reported high density of Cyanophyceae during June and July in Bhawani sagar reservoir. In the present study of Sitapai pond group cyanophyceae was numerically dominant group Bacillariophyceae. The blue green algae usually possess very efficient mechanism for intake of nutrient like nitrate, sulphate, phosphate at low concentration. Cyanophyceae occurrence can be attributed to stagnation of water, high alkalinity and nutrients.

Abundant records show that blue-green algae inhabiting deep water commonly possess a red or violet coloration. Cyanophyceae are indicators of bad quality of water. Blue green algae were also reported to be the indicator of highly polluted water (Rama Rao *et al.*. 1978).

Bacillariophyceae:

The bacillariophyceae, commonly known as the diatoms, are ever present algae of both fresh and salt water and of damp places including aerial habitats as old walls, rocky Clift's tree barks, and damp soils. They play a very important role, particularly in the aquatic vegetation planktons and the basic food of aquatic fauna. They are unicellular, sometimes colonial microscopic algae. Diatoms have been very appropriately named as the "Jewels of the plant world".

In the present study, group bacillariophyceae constituted 6.42% to 8.38% of the total phytoplankton. The minimum density of bacillariophyceae was recorded during the summer season while the maximum density of the same was observed during the winter season.

Similer observations were also made by Tripathi and Pandey (1990), Goutam Rajan *et al.*. (2007), reported high density during winter and low during summer season in Birganj, Nepal. U. Anitha *et al.*. (2007), Milind S. Hujare (2008), reported high density during winter season and low during summer season.

In the present investigation, it was observed that high dissolved oxygen and pH are favorable factors for the growth of Baccillariophyceae. Several studies have reported the effect of temperature and light on diatom community. Pearsall (1923) and Lind (1938) suggested that nitrate was the main factor controlling the periodicity of diatoms.

The chlorophyceae and Baccillariophyceae are the indicators of the healthiness ecosystem and are also responsible for the addition of high amount of oxygen to the water body, U. Anitha (2007),Brajesh and Pandey (2001).

Euglenophyceae:

The euglenophyceae represent mostly unicellular, naked motile individuals which usually occur in fresh water. Some species form dendroid colony. The Euglenophyceae, though found in less number, showed marked periodicity and abrupt disappearance.

In the present study the percentage composition of group euglenophyceae was 1.11% to 2.31%. This group is represented by two species Euglena and Trachelomonas. The maximum density was observed in the rainy season while minimum density was observed in the summer season during the study period.

Roa (1987) reported the occurrence of euglenophyceae during summer and monsoon season in Rana Sagar Lake. Kulshrestha *et al.*. (1989) reported the group in different seasons in Mansarovar reservoir in Bhopal. Choubey (1990), reported a single species of euglenophyceae in Gandhi sagar reservoir with its abundance from mid monsoon to post monsoon period. Milind S. Hujare (2008), reported high population density during rainy season and low density in the summer season in fresh water tank of Talsande, Maharashtra India.

Euglenophyceae population during rainy and winter, as observed in present study, can be attributed to high Co_2 and low dissolved O_2 which favored an abundance of euglenophyceae. This is an indication of the healthiness of ecosystem.

The results of phytoplankton study of Sitapat pond suggest that the pond is moderately polluted and showing a trend of increasing eutrophication. Ganeshan *et al.*. (2004), noticed the plankton density was uniform. Sunitha *et al.*. (2005), reported the ecological status of river ecosystem and used the phytoplankton as indicators to establish the quality status.

Table 1: easonal variation of Phytoplankton's density of Sitapat Pond (no/lit.) 20007

S. No.	Name of the group & Genera	Seasons																		Annual Total	Status
		Rainy												Summer							
		I	II	III	IV	V	VI	I	II	III	IV	V	VI	I	II	III	IV	V	VI		
Chlorophyceae																					
1	*Spirogyra* sp.	11	5	9	0	0	0	15	8	9	6	0	0	15	11	17	15	5	4	130	A
2	*Zygnema* sp.	13	3	7	3	0	0	17	10	12	5	0	0	25	19	15	9	7	2	147	A
3	*Volvox* sp.	0	0	14	16	18	11	17	10	20	22	18	15	5	8	23	12	17	11	237	A
4	*Chlorella* sp.	16	12	13	10	2	1	12	18	17	10	5	2	25	20	23	16	10	7	219	A
5	*Scenedesmus* sp.	11	8	10	0	0	0	12	11	9	7	2	1	10	17	21	12	5	0	136	A
6	*Pediastrum* sp.	0	0	12	10	18	10	0	15	17	12	15	11	13	12	18	21	17	11	212	A
7	*Ulothrix* sp.	0	0	0	11	12	11	9	17	19	10	10	12	13	22	21	12	13	7	199	A
8	*Oedogonium* sp.	2	5	13	12	17	11	10	11	18	15	4	7	10	15	20	14	19	13	216	A
9	*Cosmarium* sp.	8	5	18	10	14	3	12	10	11	9	11	18	17	15	21	25	13	10	230	A
10	*Closterium* sp.	0	0	13	15	10	11	4	6	10	11	5	8	11	17	20	21	10	11	183	A
11	*Chara* sp.	5	7	13	8	4	6	2	12	21	8	6	5	11	15	23	16	11	9	182	A
	Total species	**66**	**45**	**122**	**95**	**95**	**64**	110	**128**	**163**	**115**	76	79	**155**	**171**	**222**	**173**	**127**	**85**	**2091**	

Table 2: Seasonal variation of Phytoplankton's density of Sitapat Pond (no/lit.) 2007

S. No.	*Name of the group & Genera*	*Seasons*																		*Annual Total*	*Status*
		Rainy												*Summer*							
		I	II	III	IV	V	VI	I	II	III	IV	V	VI	I	II	III	IV	V	VI		
Cynophyceae																					
1	*Oscillatoria* sp.	9	11	10	9	0	0	10	11	12	13	0	0	15	16	13	18	8	1	156	A
2	*Anabaena* sp.	0	0	12	14	8	10	0	0	11	15	10	11	5	7	11	14	12	11	151	A
3	*Nostoc* sp.	0	0	15	10	8	7	0	0	15	18	6	0	4	6	12	15	10	12	138	A
4	*Merismopedia* sp.	3	5	11	12	15	3	5	2	11	15	8	0	7	9	11	12	10	8	147	A
5	*Microcystis* sp.	0	10	15	18	10	4	7	8	18	19	7	3	6	10	17	15	10	11	188	A
6	*Spirulina* sp.	5	0	10	12	9	10	5	9	18	19	15	8	2	10	15	18	16	12	193	A
	Total species	17	26	73	75	50	34	27	30	85	99	46	22	39	58	79	92	66	55	973	
Bacillariophyceae																					
1	*Melosiras* sp.	0	0	12	18	15	21	5	7	19	15	15	16	2	3	15	10	8	2	183	A
2	*Fragilaria* sp.	0	19	15	12	13	8	11	18	17	10	9	13	2	23	5	7	10	0	192	A
3	*Navicula* sp.	4	6	19	17	15	2	10	12	11	11	12	5	0	0	13	15	17	11	180	A
4	*Pinnularia* sp.	5	8	11	18	14	12	8	12	11	10	11	9	11	5	12	15	12	13	197	A
5	*Amphora* sp.	0	0	11	18	9	6	4	9	12	13	15	13	0	0	4	5	7	3	129	A
	Total species	9	33	68	83	66	49	38	58	70	59	62	56	15	31	49	52	54	29	881	
Euglenophyceae																					
1	*Euglena* sp.	3	12	19	15	2	1	0	10	15	12	0	0	0	8	7	9	0	0	113	A
2	*Trachelomonas* sp.	6	5	16	13	0	0	0	6	8	9	0	0	0	2	5	6	0	0	76	C
	Total species	9	17	35	28	2	1	0	16	23	21	0	0	0	10	12	15	0	0	189	

A=Abundence,C=Common

Table 3: Seasonal variation of Phytoplankton's density of Sitapat Pond (no/lit.) 2008.

S. No.	*Name of the group & Genera*	*Seasons*																		*Annual Total*	*Status*
		Rainy						*Winter*						*Summer*							
		I	**II**	**III**	**IV**	**V**	**VI**	**I**	**II**	**III**	**IV**	**V**	**VI**	**I**	**II**	**III**	**IV**	**V**	**VI**		
Chlorophyceae																					
1	*Spirogyra* sp.	7	8	10	0	0	0	17	7	10	5	2	0	14	12	15	17	3	2	129	A
2	*Zygnema* sp.	10	6	11	2	0	0	15	11	13	6	0	0	11	12	21	10	5	0	133	A
3	*Volvox* sp.	0	0	15	12	10	11	10	15	18	17	13	11	8	10	17	15	11	9	202	A
4	*Chlorella* sp.	15	12	13	11	0	0	11	12	17	13	6	3	15	17	15	16	5	2	183	A
5	*Scenedesmus* sp.	11	8	9	0	0	0	13	12	8	7	0	0	11	17	18	9	5	1	129	A
6	*Pediastrum* sp.	0	0	12	11	15	12	0	15	17	13	10	1	11	13	17	12	9	2	170	A
7	*Ulothrix* sp.	0	0	3	12	13	10	11	12	15	18	5	2	10	11	17	8	7	5	159	A
8	*Oedogonium* sp.	0	5	16	15	11	2	10	11	13	17	4	9	12	15	18	13	9	4	184	A
9	*Cosmarium* sp.	11	10	12	9	7	2	9	21	19	12	10	7	12	10	15	17	6	3	192	A
10	*Closterium* sp.	0	0	10	9	18	16	5	7	11	16	14	15	12	10	12	15	12	9	191	A
11	*Chara* sp.	5	8	9	10	5	4	9	11	12	9	5	4	11	17	12	15	9	5	160	A
	Total species	**59**	**57**	**120**	**91**	**79**	**57**	**110**	**134**	**153**	**133**	**69**	**52**	**127**	**144**	**177**	**147**	**81**	**42**	**1832**	

Table 4: Seasonal variation of Phytoplankton's density of Sitapat Pond (no/lit.) 2008

S. No.	*Name of the group & Genera*	*Seasons*																		*Annual Total*	*Status*
		Rainy						*Winter*						*Summer*							
		I	**II**	**III**	**IV**	**V**	**VI**	**I**	**II**	**III**	**IV**	**V**	**VI**	**I**	**II**	**III**	**IV**	**V**	**VI**		
Cynophyceae																					
1	*Oscillatoria* sp.	0	2	16	8	0	0	5	2	12	15	0	0	4	5	8	18	11	11	117	A
2	*Anabaena* sp.	0	0	15	16	8	10	0	0	10	13	11	10	7	8	11	13	9	7	148	A
3	*Nostoc* sp.	5	3	11	13	12	3	6	4	12	15	8	4	9	8	13	14	11	12	163	A
4	*Merismopedia* sp.	10	8	13	12	10	5	4	6	11	13	9	3	7	11	15	16	10	5	168	A
5	*Microcystis* sp.	0	0	13	9	8	7	5	7	12	15	16	9	13	10	16	15	9	3	167	A
6	*Spirulina* sp.	2	3	11	18	17	7	5	10	13	15	10	8	10	11	17	15	16	5	193	A
	Total species	17	16	79	**76**	**55**	**32**	**25**	**29**	**70**	**86**	**54**	**34**	**50**	**53**	**80**	**91**	**66**	**43**	**956**	
Bacillariophyceae																					
1	*Melosiras* sp.	2	3	12	15	10	12	6	7	15	14	18	12	0	0	13	16	15	11	181	A
2	*Fragilaria* sp.	0	0	10	15	13	10	8	11	10	12	17	9	0	13	14	15	14	11	182	A
3	*Navicula* sp.	0	0	11	17	13	15	0	0	11	15	19	10	5	7	15	11	18	12	179	A
4	*Pinnularia* sp.	7	8	15	10	8	6	5	6	15	18	5	8	7	9	13	11	21	13	185	A
5	*Amphora* sp.	0	0	10	11	18	11	7	10	18	16	19	7	0	0	5	4	7	4	147	A
	Total species	9	11	58	68	62	54	26	**34**	**69**	**75**	**78**	**46**	**12**	**29**	**60**	**57**	**75**	**51**	**874**	
Euglenophyceae																					
1	*Euglena* sp.	10	11	12	17	2	2	10	0	12	13	0	0	0	0	12	10	0	0	111	A
2	*Trachelomonas* sp.	2	12	10	15	0	0	5	9	13	10	0	0	9	8	6	7	0	0	106	A
	Total species	12	23	22	32	2	2	15	9	25	23	0	0	9	8	18	17	0	**0**	**217**	**A**
A=Abundance																					

Table 3: Yearly veriation of Phytoplankton density in Sitapat pond (mean+SD×10^2)

S.No.	*Name of Genera*	*Year 2007*	*Year 2008*
1	Chlorophyceae	53.14 ± 3.000	54.23 ± 2.867
2	Bacillariophyceae	20.68 ± 0.673	21.31 ± 1.434
3	Cynophyccae	19.98 ± 2.645	19.21± 3.506
4	Euglenophyceae	5.16 ± 0.990	5.12 ± 0.659
Total		99.66 ± 7.308	99.87 ± 8.466

REFERENCES

1. Ayyappan, S. & Jena J. K. (2003): Grow-out production of Carps in India. Journal of Applied Aquaculture, 13: 251-282.
2. Clarke, L.R. and D.H. Bennett. (2003): Seasonal zooplankton abundance and size fluctuations across spatial scales in Lake Pend Oreille, Idaho Journal of fresh water Ecology 18 (2): 333-336
3. Das, S.K. and B.K. Chand. (2003): Limnology and biodiversity of Icthio fauna In a pond of southern Orissa, India. J.E. Ecotoreicol. Environ. Mohit. 13(2): 97-102
4. Milind, S. Hujare (2008): Limnological studies of the perennial water body, Attigre tank, Kolhapur District (Maharashtra). Nature, Environment and pollution Technology: vol 7(1): 43-48.
5. Pathak, S.K. and L.K. Mudgal (2005): Limnology and biodiversity of fish fauna in Virla reservoir, M.P. India.
6. Patil, H.S. and S.M. Karikal (2001): Zooplankton diversity of Bhutnal reservoir at Bijapur- Karnataka state. In water quality assessment biomonitoring and zooplankton diversity, (Ed.) B.K. Sharma. PP.236-249.
7. Pawar, S., Sharma, D.K. and Mishra R. (2001): Physico chemical and Microbiological aspects of water pollution in cheelar reservoir, Shajapur. Abstract National sem., Khargone: 34.

8. Prasad and Singh (2003): Composition, abundance and distribution of Phytoplankton and Zoobenthos in a tropical water body. Nat. Envin. Pollut. Technol. 2: 255-258.

9. Raveen, R.C. Chenna Krishnan and A. Stophen (2008): Impact of pollution on the quality of water in three fresh water lakes of sub-urvon chennai; Nature, Environment and pollution Technology; Vol. 7(1). PP. 61-64.

10. Sharma, Archana (2007): Hydrobiological studies of Yeshwant sagar reservoir with special reference to commercially important fishes, Indore (M.P.). A Ph.D. Thesis, D.A.V.V., Indore, India

11. Welch, P.S. (1952): Limnology. McGraw Hill Book Co., New York. P.538.

Floral Diversity and their Conservation (2013), Editors: D.R. Khanna et al.
Published by Biotech Books.
ISBN: 978-81-7622-286-0 Pages: 191-196

20

Effect of Seed Treatment with Eco-friendly Non-toxic Plant Origin Substances on Seed Quality Paramcters of Tur during Storage

SHANTI PATIL[1], SANJEEV CHARJAN[1], RAJESH GADEWAR[2], ASHISH LAMBAT[2]
[1]Assistant Professor, Dr. P D K V's, College of Agriculture, Nagpur, M.S, India
[2] Assistant Professor, Sevadal Mahila Mahavidyalaya, Nagpur, M.S, India.

A laboratory experiment was conducted to study the effect of different doses (1, 1.5 and 2%) of *Acorus calamus* rhizome and Croton tiglimum (seed) powder on seed quality *viz.*, 100-seed weight, germinability and vigour index of tur (*Cajanas cajan* L.) seeds and their efficacy against the development of store grain pest, pulse beetle (*Callosobruchus Chinensis Fab.*) during storage. Pre-storage seed treatment with *Acorus calamus* rhizome powder and Croton tiglimum seed powder in the proportion of 2.0% by seed weight was found most effective in maintaining better seed quality parameters and arresting the development and infestation of store grain pest, pulse beetle during storage.

Key words: Eco-friendlv. *Acorus calamus*, croton tinlium seed powder, seed quality, tur.

INTRODUCTION

Tur (*Cajanus cajan* cv. c-11), a Kharif Crop, occupies an exalted position amongst the protein rich bean, being widely used as pulse as well as vegetable. Harvested seeds were stored for varying periods before consumption or sowing purposes. Tur seed was found to be suitable host to pulse beetle during storage (Yadva, 1985). Its infestation, either originates in the field or in storage, causes loss to the seeds.

Pre-storage seed treatment with some sort of protectant to avoid pest infestation during storage has an old age practice. Saxena *et al.* (1976) and Tikku *et al.* (1978) found *Acorus calamus* L. oil vapour responsible for causing infecundity among the females of a number of stored grain pests. The objective of the present investigation was to study the effect of different doses of *Acorus calamus* (rhizome) and *Croton tiglimum* (seed) powder on seed quality of tur and their bioefficacy against the development of pulse beetle during storage.

MATERIALS AND METHODS

Tur (*Cajanus cajan* cv. c-11), seeds were used in various phases of this study, produced in 2010 (Kharif Season). The seeds were cleaned and dried (Moisture content 10.7%). The seeds were treated (Jan. 2011) with two plant products *viz. Acorus calamus* (rhizome) and *Croton tiglimum* (seed) powder each in the proportion of 0.2, 0.4 and 0.6% by weight of seed. The experiment as conducted in glass bottle of one litre capacity with seven treatments including untreated control. Each glass bottle was then filled with 500 grams of tur seeds. Ten pairs of 2-3 days old pulse beetle (*Callosobruchus analis* Fab.) maintained at 25 ± 2^{0}C on tur seeds having 14% moisture at ambient temperature and relative humidity were released in each glass bottle covered with muslin cloth. The set of experiment was kept in well ventilated wire mesh almirah in mesonary building having cemented walls, roof and floor under ambient temperature (10.8-37.2^{0}c) and relative humidity (49.7-85.1%) from January to March 2011. After 3 months the grains from each treatment were keenly observed and those found infested were separated out weighed to determine the infestation percentage on weight basis, 100-seed weight and germination were tested in quadruplicate with100-seeds in each replication. The germination percentage was evaluated on the value for percent normal seedlings (Anonymous, 1985). The vigour index was worked out following the method of Abdul-Baki and

Anderson (1973).The data were statistically analyzed in RBD using Panse and Sukhatme (1967).

RESULTS AND DISCUSSION

The data regarding the effect of the different doses of *A. calamus* (rhizome) and *C. tiglimum* (seed) powder on population behaviour of pulse beetle, infestation percentage, 100-seed weight, germination percentage and vigour index of tur seed after 3 months of storage are given in Table 1.

Table 1: The effect of different doses of *A. calamus* (rhizome) and *C. tiglimum* (seed) powder on seed quality of tur and their bioefficacy against pulse beetle during storage.

S. No.	*Treatments*	*Conc.*	*Increase in number of insects 500 g seed*	*Percentage Infestation (W/W)*	*100-seed weight (g)*	*Germination percentage*	*Vigour Index*
1	*A. calamu*	1.0%	8.02	2.06	9.62	76.24	1842
2	*A. calamu*	1.5%	5.52	1.26	9.96	81.80	1912
3	*A. calamu*	2.0%	2.78	0.25	10.26	85.02	2289
4	*C. tiglimum*	1.0%	9.84	2.49	9.22	74.06	1806
5	*C. tiglimum*	1.5%	6.82	1.39	9.54	79.20	1814
6	C. tiglimum	2.0%	3.04	0.52	9.72	82.44	1896
7	*Untreated Control*		92.52	24.01	6.02	21.92	492
	SE ±		14.16	1.44	0.04	0.82	--
	CD at 5%		43.49	4.52	0.13	2.44	--

The results indicated the variation in number of pulse beetle (adult) in each treatment. Minimum population of pulse beetles was observed in treatment of *A. calamus* (rhizome) and *C. tiglimum* (seed) powder in the proportion of 2% by weight loss. The number of pulse beetle decreased with the increase in concentration. The active ingredient of the *A. calamus* (rhizome) and *C. tiglimum* (seed) may be responsible in arresting the developmental activities or mortality of the pulse beetle. The untreated control showed maximum population of pulse beetles during storage. The results are in agreement to those reported by Charjan (2011). *Acrorus calamus* (rhizome) oil vapour caused infecundity in the terminal follicle of the vitellarium among the females at the pulse beetle (Saxena *et al*, 1976). Tikku *et al*. (1978) concluded from their studies that *A. calamus* oil has got sterilizing effect on the female of *Trogoderma granarium* Evert. The results reported here are in conformity with the observations made by earlier workers Khan and Borle (1985) and Charjan and Tarar (1994).

The data indicated that the infestation percentage of pulse beetle declined with the increase in concentration of *A. calamus* (rhizome) and *C. tiglimum* (seed) powder. A sharp declined in infestation percentage occurred in seed treated with 2% concentration of *A. calamus* (rhizome) and *C. tiglimum* (seed) powder. The untreated seed showed highest percentage of infestation. This might be due to the increase in population of pulse beetles. The results are in agreement to those report by Khan and Borle (1985) and Charjan and Tarar (1994).

The seed quality parameters *viz.* 100-seed weight, germination percentage and vigour index follow the same tend of infestation percentage. 100-seed weight, germination percentage and vigour index was highest in seed treated with 2% concentration of *A. calamus* (rhizome) and *C. tiglimum* (seed) powder. This might be due to the least infestation of pulse beetles during storage. The 100-seed weight, germination percentage and vigour index decreased with increasing infestation of pulse beetle (Howe, 1972). Since the pulse beetle have been eaten off major portion of the endosperm which leads to reduction in weight of the seed and inturn affects the seed germination and vigour index because of lack of stored food and is in conformity with the findings of Narayanaswamy (1985). Charjan and Tarar (1994), Charjan *et al*. (2006) and Wankhede *et al*. (2010). Handerson and Christensen (1961) reported that pulse beetle attack the embryo and germination potential of seed reduced or totally destroyed.

Khan and Borle (1985) and Charjan (1993) reported that the storability of pulse seed could be prolonged without infesting pulse beetle by treated seed with *A. calamus* (rhizome) and *C. tiglimum* (seed) powder. Khan *et al*. (1982) recorded minimum storage losses in stored sorghum when seed treated with *A. calamus* rhizome powder @ 0.1% coupled with burning camphor for oxygen deconcentration. Thus it can be concluded from the result that tur seeds treated with *A. calamus* (rhizome) and *C. tiglimum* (seed) powder in the proportion of 2% can maintain good seed quality and check the population behaviour of pulse beetle during storage.

REFERENCES

1. Abdul-Baki, A.A. and Anderson, J. D. 1973. Vigour determination in soybean seed by multiple criteria. Crop. Sci. 13: 630-633.
2. Anonymous, 1985. International rules for seed testing. Seed Sci. and Technol. 13: 299-513.
3. Charjan, S.K.U. 1993. Effect of storage conditions on storability of gram (*Cicer arietinum* L.) Paper presented at conf. on Recent Trends in Botany held at Amravati University on 6-8 Dec. 1993.
4. Charjan, S.K.U. and Tarar, J. L. 1994. The influence of some plant products on seed quality of lobia during storage. Ann. Plant Physiol. 8(2): 153-156.
5. Charjan, S.K.U., Wankhede, S. R. and Jayade, K. G. 2006. Viability, vigour and microflora changes in pulse beetle damage seeds of cowpea produce in dryland condition. Proceeding of Economics of Sustainability of Dryland Agriculture, p.5.
6. Charjan, S.K.U. 2011. Studies on the influence of *Acorus calamus* L. rhizome powder seed treatment against store grain pest of wheat. Proc. of National Seminar on environmental management and biodiversity Conservation, Rishikesh, Abstr. No.127, p.101.
7. Handerson, L. S. and Christensen C. M. 1961. Preharvest control of insect and fungi. U.S. Dept. Agri.Ybk, pp.348-356.
8. Howe, R. W. 1972. Insect attacking seed during storage. In: Seed Bioloty Vol. III (ed. Kozlowski, T. T.) Academic Press, New York, pp.247-300.
9. Khan, M. I., Rajurkar, B. S. and Borle, M. N. 1982. Assessment of storage losses due to insect pests in grain sorghum treated with non-toxic plant origin substances. P.K.V. Res. J. 6(1): 45-48.
10. Khan, M. I. and Borle, M. N. 1985. Efficacy of some safer grain protectants against the pulse beelte, Callosobruchus chinesis L. infesting stored begal gram (*Cicer arietinum* L.) P.K.V. Res. J. 9(1): 53-55.
11. Narayanaswamy, S. 1985. Effect of pulse beetle damage on seed quality of field bean and pigeonpea. Seed Res. 13(2): 138-141.
12. Panse, V. G. and P. V. Sukhatme. 1967. Statistical methods for agril. Workers I.C.A.R. Pub. New Delhi.
13. Saxena, B. P. Koul, O. and Tikku, K. 1976. Non-Toxic protectants against the stored grain insect pests. Bull. Grain Technol. 14(5): 190-193.

14. Tikku, K., Koul, O. and Saxena, B. P. 1978. The influence of Acorus calamus L. oil vapour on the histocytological pattern of the ovaries of Trogoderma granarium Evert. Bull Grain Technol. 16(1): 3-9.
15. Wankhede, S. R. Gajbhiye and Charjan, S. U. 2010. Studies on the quality of lobia seeds influenced by the infestation of pulse beetle during storage. Green Farming. 1(6): 654-655.
16. Yadav, T. D. 1985. Host suitability of different legume seed to important pulse beetles. Seed Res. 13(2): 29-34.

Floral Diversity and their Conservation (2013), Editors: D.R. Khanna et al.
Published by Biotech Books.
ISBN: 978-81-7622-286-0 Pages: 197-209

21

Biodiversity of Nahergarh Hill Forest nearby Mansager Reservoir (Jalmahal) with Special Reference to Nature based Ecotourism

P.K. Sharma, K.P. Sharma and Subhsini Sharma*
Department of Botany, University of Rajasthan, Jaipur, India
*Department of Zoology, University of Rajasthan, Jaipur, India

Nature trails and eco-tourism parks have been developed as an attempt to reverse the damage done by territorial encroachment and to recreate a pristine environment in Rajasthan. There are several at a distance of less than 10 km in Jaipur.

The vegetation of the study area is represented through tropical dry deciduous and tropical thorn forests having three strata. The important tree of the top layer are; *Anogeissus pendula*, *Acacia senegal*, *Boswellia serrata*, *Butea monosperma* and *Holoptelea integrifolia*. Important plant species of the second tier are *Adhatoda vasica*, *Capparis sepiaria*, *Indigofera tinctoria*, *Rhus mysurensis*, *Grewia tenax*, *Euphorbia neriifolia* and *Opuntia elatior*. The forest floor is covered with green carpet of herbaceous vegetation (3rd tier) that last until rains are over. The senescence of ground flora beginning in the post-rainy season is over by onset of winter (November). The forest floor is now visible again, though littered.

Thick forest cover on Aravallis makes study area an ideal home for wild life. Twenty one species of wild mammals have been recorded by the State Forest Department and many of them had been sighted (15) during my field visits of the study area. Of great interest for tourists is Common Indian Monitor sighted frequently during summer in the study area. Mansagar lake supports rich avian diversity of both terrestrial and waterfowl populations. In all 92 species of birds were sighted during the study period. Avian diversity increased markedly during winter after arrival of migratory birds. During the present study, Mansagar Lake & its Environ has been explored in detail with a particular reference to nature based tourism, also termed as Ecotourism. The forest along Mansagar lake is rich in fauna. Sacred groves playing greater role in conserving floral and faunal diversity Mansagar Environ are; Shree Vir Hanuman temple, Sita Ram temple and Annapurna devi temple.

Key words: Nahargarh hill forest, Ecotourism, Avian diversity, Flora and Fauna

INTRODUCTION

Interestingly, the concepts of ecotourism and biodiversity conservation have evolved at roughly the same time. Numerous opportunities and benefits can be derived by strategically integrating biodiversity conservation requirements with future tourism needs. The goal of ecotourism is to capture a portion of the enormous global tourism market by attracting visitors to natural areas and using the revenues to fund local conservation and fuel economic development. Jaipur, "Pink City of India" is a heritage city forming part of "Golden Triangle" formed between Agra – Delhi – Jaipur. Every second international tourist visiting India stays in Jaipur. The hills have good forest cover that provides shelter to a variety of wild life.

STUDY AREA

The founder, Sawai Jai Singh II constructed Jaipur in the year 1727 on a sandy flat plain in the corner where the two anticline ranges of the Aravallis, the Galta-Jhalana on the east and the Nahargarh on the north, converge. Their average height is of 500 meters and the highest peak of Jhalana range is 589 meters and in the Nahargarh 587 meter. Mansagar lake is a large manmade lake on the northern fringe of Jaipur city at about 5 km from the city center on the way to Amer. The lake is approximately 130 ha in its spread with a maximum depth (> 7m) at the dam's outflow point. It has a catchments of 23.5 km2, about 40% of which falls inside the dense urban area is presently the major source of water (about 90%) in the lake in the form of storm water runoff through Nagtalai and Brahmpuri Nalas in the rainy season while sewage (until 2007) in the dry weather.

MATERIALS AND METHODS

The regular field visits were made in the forest to collect plant specimens. Their frequencies were higher during rainy season since most of the herbaceous ground flora is of ephemeral nature. The plant specimens were brought to the laboratory and their herbarium sheets were prepared and were matched with sheets in the Herbarium, Department of Botany, UOR,Jaipur.The plant identification was made in consultation of herbarium curator Lt. Sh. Roop Singh using standard floras (Sharma and Tiagi 1979, Bhandari 1990, Sheety and Singh 1987- 93). The animals sighted during field visits were identified with the assistance of forest guards posted there for almost 5 years. They are experienced persons working with Forest Rangers for annual census of wild life in the forest. Besides, a check list of animals in the forest prepared by the Rajasthan State Forest Department was consulted. The specimens of fish caught by the contractor (2005-06) from Mansagar lake were procured and identified by my Co-supervisor Dr. Subhasini Sharma.The vertebrate fauna, especially avian, were identified with the help of standard monographs (Ali 1974, Cruickshank 1960, Kazmierczak and Perlo 2000, Kumar *et al.*. 2005). Their habitats, counts and feeding habits were studied during the course of study. Based on 2 years data birds are classified into resident and migratory in the study area. The nesting of dominant resident birds was monitored. Insects and butterflies caught by trap were identified using standard monograph (Tonapi 1980).

RESULT AND DISCUSSION

The hills have a good forest cover is represented through tropical dry deciduous and tropical thorn forests. It has three strata. The top layer comprises of 30-40'tall trees, middle 6-12' long shrubs and under shrubs while the lowest is the herbaceous layer (2-5'). The forest had 14 woody species, *Anogeissus* being the key stone species on the middle and top hill sites of the forest while *Prosopis juliflora* on the foot hill. Litter fall contributes nutrients to soil in winter but practice of collecting *Anogeissus* litter in winter as fodder by villagers is detrimental to forest ecosystem. It also affects food security of wild herbivores in the forest, for them leaf litter constitutes major supply of food.

The important tree species of the top layer are; *Anogeissus pendula*, *Acacia senegal*, *Boswellia serrata*, *Butea monosperma* and *Holoptelea integrifolia*. Important plant species of the second tier are *Adhatoda vasica*, *Capparis sepiaria*, *Indigofera tinctoria*, *Rhus mysurensis*, *Grewia tenax*, *Euphorbia neriifolia* and *Opuntia elatior*. During rainy season, an extra 4th tier comprising of thick mats of *Riccia* and *Adiantum* was noted in moist shady areas. Grasses like *Digitaria pennata* grows in between shrubs. Immediately after onset of rains, hills turn green. The forest floor is covered with green carpet of herbaceous vegetation that last until rains are over. The senescence of ground flora beginning in the post-rainy season is over by onset of winter (November). The forest floor is now visible again, though littered.

Anogeissus pendula is the key stone species of Aravallis hill ecosystem at Jaipur. The charcoal made from its wood has high calorific value- a property that led to its over exploitation in Rajasthan including Jaipur in the recent past. It however, grows luxuriantly in the forest falling under study area. I have observed its regeneration through seeds in the vicinal areas of Sitaram temple. The tree changes its color from green in the rainy season to ash in the winter, and sc, also hill view seasonally. The leaves finally fell down in the winter. Being from a leguminous family, these are rich in protein, and therefore, a nutritive food to forest herbivores during winter when they don't have access to herbaceous species. It is to be noted here that herbaceous ground flora of the forest is grazed by that time resulting in food scarcity for herbivore. Thus, *Anogeissus* leaves are the nature gift to wild herbivore. Unfortunate, man is competing with the wild population, as he collects dried leaves from the forest floor during winter. Similar collection of herbaceous ground flora during rainy season may not harm wild population since it grows abundantly. But this practice has to be curbed for sustaining good grazing food chains in the forest ecosystems to enrich wild life- a key element in ecotourism.

Acacia senegal, *Boswellia serrata*, *Butea monosperma*, *Holoptelea integrifolia* and *Sterculia urens* are the other tree species noted occasionally in the forest.

Acacia senegal growing at the middle of Aravallis looks attractive during flowering and fruiting. Its pods are used in preparation of traditional preparation "Panchkuta" while gum is of medicinal value. It is seen growing near Sitaram temple.

Boswellia serata, commonly termed as sentinel of Aravallis, grows on the top of the hill. It flowers in the post-rainy season forming a distinct yellowish golden belt in the green background pleasing to eyes. All leaves are lost during winter leaving silvery leafless shoots shining in the sun. Its gum is also important medicinally.

Butea monosperma also forms a distinct belt in the middle of hills in the Mansagar environ which is very conspicuous during summer since it is the only green tree in the ash colored background of *Anogeissus*. Its bright red flowers in spring form a garland around the hill, when watched closely. Increasing the number of its trees by planting sapling will certainly improve landscape as we have done on the lake bank. Its flowers yield a natural dye which is in great demand in the textile industry. Onsite demonstration of dye application to fabric may fetch appreciation to guide by tourists. It is said that this dye was used by the lord Krishna on the occasion of holy – the most popular Hindu festival celebrated during spring season. The leaves of the plant are in used for preparing bowls and trays used during traditional feast – a good example of hygiene in Indian feast to be explained to tourists. The nearby restaurants may be encouraged to offer snacks in such utensils, on tourist request.

Holoptelia integrifolia, commonly known as Monkey Bread, grows at the foot hill. It is a favorite destination for roosting of waterfowls in the Mansagar lake. Fruiting is profuse in the tree which after dispersal is consumed by children in the rural areas. The guide may encourage tourists to taste fruits as part of fun during their visit into wilder areas.

Adhatoda vasica, commonly known as Vanasfa, is a common perennial under-shrub forming middle strata of the forest. The decoction of its leaves is used in cold, cough and bronchitis by the rural folks - this knowledge may be shared with a tourist to demonstrate application of local biodiversity in medicine.

Capparis sepiaria is a threatened shrub is of great medicinal value. It grows in the middle of Aravallis in Mansagar environ, being more common in the forest area nearby Sitaram temple. In Nahargarh Biological Park, it grows in plain.

Commiphora wightii is a slow growing **endangered** species of Aravallis. Single naturally growing specimen has been noticed in the Sitaram temple since 2005. One more specimen has been observed in August 2008 along the lake bank. I have planted 50 saplings of this species along the lake bank in 2008. It is a medicinally important plant.

Euphorbia neriifolia grows abundantly both in the middle and top of the Aravalli hills. It is a laticiferous plant yielding products similar to that of petroleum after extraction. This plant species has been explored by scientists in India for renewable source of energy and may be of interest to students on excursion in the forest.

Opuntia elatior grows with *Euphorbia* in the middle of Aravallis. In our visit to World Forestry Arboretum, near Jhalana hill, a scientist from Brazil was amazed to see an insect feeding on this plant. He picked up the insect and crushed between fingers that released a dark red dye in plenty which is used for printing in Brazil. He informed us that cultivation practice has been developed for this insect using *Opuntia* as feed for commercial production of natural dye in great demand in the developed world. It's a very expensive dye (about $ 1000/kg). Our state may also develop its cultivation practice for employment generation in rural areas where *Opuntia* is available in plenty.

Grewia tenax and *Rhus mysurensis* are shrubs growing in the foot and middle of Aravallis hills. The fruits of *Rhus* are sold in market during rainy season as children are fond of them. The guide may also pursue tourists to taste wild edible of India.

Indigofera tinctoria and *Jatropha gossifolia* were observed only in the foot hills adjoining Mansagar lake.

Ficus glomerata and *Mitragyna parvifolia* grows in the marsh of Annapurna devi complex. These tree species support food chains dominated by avian.

Phoenix sylvestris is the dominant species in the Annapurna temple marsh. Its presence improves landscape while fruits are dear to several birds species.

The herbaceous species of interest to biology students are *Riccia* (bryophyte) and *Adiantum* commonly known as walking fern (Pteridophyte).

Both these are taught in undergraduate classes to explain evolution in the Plant Kingdom. They are also element of seral stages in succession – a process explaining ecosystem development.

The important macrophytes noted along the bank of water bodies and temple marsh (Annapurna devi temple complex) are *Bacopa monnieri*, *Cyperus alopecuroides*, *Eclipta prostrata*, *Lemna aequinoctialis*, *Spirodela polyrhiza*, *Pandanus odoratissimus*, *Paspalum* Sp., *Paspalidium* Sp., *Phragmites karka*, *Polygonum plebeium*, *Polygonum barbatum*, *Ranunculus scleratus*, *Scirpus* sp., *Typha angustata* and *Wolffia arhiza.*

Wolffia arhiza is the smallest flowering plant noted for the first time in check dam in November 2008. It may be of interest to people visiting lake and biology students in particular. It is commonly known as Watermeal, one of the Duckweeds, in the family Lemnaceae. It is not considered invasive and might actually have economic value. It has approximately the same percentage of protein as soy beans.

Bacopa monnieri, commonly known as Brahmi, is another important marshy species of greater interest to tourists. It is the source of a drug Memory Plus developed by Central Drug Research Institute, Lucknow. Also recent studies showed that *Bacopa* protects cells against Al toxicity (Jyoti *et al.*. 2007).

40 species of grasses and forbs were recoded in the herbaceous vegetation in forest on Nahargargh hills (Aravalli hills) nearby Mansagar lake. *Achyranthes aspera, Brachiaria ramosa, Blainvillea acmella, Catharanthus pusila, Cassia absus, Commelina undulata, Hibiscus lobatus, Leucas aspera Peristrophe bicalyculata* and *Triumfetta rhomboidea* were the dominant species of vegetation in the rainy season.

Pandanus odoratissimus is the unique marshy species noted in the temple marsh. Once forming a large complex in the marsh, it has now been confined into two small pockets. During flowering in the rainy season, the surroundings are scented with a mild fragrance.

The visit to temple marsh revealed a unique practice of harvesting marshy species for green fodder for more than 50 years, as per available records. This is a good example to educate common people about wetlands value for mankind, besides serving as kidney of earth. The harvesting also scavenges nutrients from the polluted water – a subject matter for ecology students.

Thick forest cover on Aravallis is an ideal home for wild life. The forest along Mansagar lake support good fauna, especially Indian hare (about 30-50), Common Indian Monitor (about 150) and Mongoose (about 50). Other wild life counts reported by forest guard are; panther (2-3), stripped hyena (1), dear (5-6), sambhar (3-5) and nilgai (2).

In 2007,a female panther with her two cubs was sighted during winter. Similarly, one nilgai cub trapped in marshy area of the lake was rescued in September 2007. Thus, forest supports good food chains to be favorable for their multiplication.

Among reptiles, of great interest to tourists is Common Indian Monitor (number = 150) seen commonly during summer. Thick plantation along the lake has increased food availability to rodents and so, of snake population, especially of Indian Cobra seen frequently. It is not out of place to mention here that both Grey and Small Mongoose breed here. Thus, a perfect food chain has established.

Other important animals interesting to both tourists and biology students visiting lake for excursions are butterflies, spiders, aquatic insects . The dragon fly (biological control of mosquito) also known as helicopter (all the four wings move independently and so insects remained idle during flight) is of great interest. This principle has been used to design Chauffer. Thus, these insects are the live demonstration of scientific principle to students. The State Forest Department has made a walk way around Check dam to watch waterfowls.

In all 92 species of birds were sighted in the study area, especially Mansagar lake environ, avian diversity increased markedly during winter due to arrival of migratory birds in the lake. The birds inhabiting wetlands for nesting, feeding and roosting are broadly defined as water birds. They play a significant role in human lives culturally, socially and scientifically and as a food resource, besides being ideal indicators of wetland health. These birds have attracted man since time immemorial and find mention in many ancient epics. Wetland birds form vital links in the food webs. They often occur in amazing numbers on wetlands and are the best indicators of the biodiversity richness of these productive ecosystems. The water birds are probably the most prominent groups which attract people to wetlands.

Table 1: Bird populations recorded in the forest area, including bank of Mansagar reservoir during May 2005 –Dec.2007.

S. No.	*Common Name & Scientific Name*	*Niche*
1.	White Breasted Kingfisher: *Halcyon smyrnensis*	Sb, F
2.	Blue Rock Pigeon: *Columba livia*	F
3.	Eurasian (Ring) Dove: *Streptopelia decaocto*	F
4.	Laughing Dove: *Streptopelia senegalensis*	F
5.	Asian Koel: *Eudynamys scolopacea*	F
6.	Spotted Owlet : *Athene brama*	F
7.	Green Bee-Eater: *Merops orientalis*	F
8.	Black Drongo: *Dicrurus adsimilis*	F
9.	Comman Myna: *Acridotheres tristis*	F
10.	Asian Pied Starling: *Sturnus contra*	F
11.	Rufous Tree Pie: *Dendrocitta vagabunda*	F
12.	House Crow: *Corvus splendens*	F
13.	Jungle Babbler: *Turdoides striatus*	F
14.	House Sparrow: *Passer domesticus*	F
15.	Oriental Honey –Buzzard: *Pernis ptilorbynchus*	F
16.	Grey Francolin: *Francolinus pondicerianus*	F
17.	Indian Peafowl: *Pavo cristatus*	F , Sb
18.	Greater Coucal:*Centropus sinensis*	Sb, F
19.	Blue Tailed Bee Eater: *Merops philippinus*	F
20.	Grey-Backed Shrike: *Lanius tephronotus*	F
21.	Great Grey Shrike: *Lanius excubitor*	F
22.	Red-Vented Bulbul: *Pycnonotus cafer*	F
23.	Plain Prinia: *Prinia inornata*	F
24.	Jungle Prinia: *Prinia sylvatica*	F
25.	Bluethroat Robin: *Luscinia svecica*	F
26.	Indian Robin: *Saxicoloides fulicata*	F
27.	Purple Sunbird: *Nectarinia asiatica*	F
28.	Pariah Kite: *Milvus migrans*	F
29.	Brahminy Kite: *Holiastur indus*	F
30.	Brown Capped Woodpecker: *Dendrocopos nanus*	F
31.	White-Browed Fantail: *Phipidura aureola*	F

contd...

S. No.	*Common Name & Scientific Name*	*Niche*
32.	Tree Pipate: *Anthus trivialis*	F
33.	Great Tit: *Parus major*	F
34.	Brahminy Starling: *Sturnus pagodarum*	F
35.	Common Stonechat: *Saxicola torquata*	F
36.	White Bellied Drongo: *Dicrurus caerulescens*	F
37.	Chestnut Shouldered Petronia: *Petronia xanthocollis*	F
38.	Black Red Start: *Phoenicurus hodgsoni*	F
39.	Brown Rock-Chat: *Cercomela fusca*	F
40.	Small Minivet: *Pericrocotus cinnamomeus*	F
41.	Common Hoopoe: *Upupa epops*	F
42.	Rose-Ringed Parakeet: *Psitacula krameri*	F
43.	Plum-Headed Parakeet: *Psitacula cyamocephala*	F
44.	Indian Roller; *Coracias bengbalensis*	Sb, F
45.	Pied Cuckoo; *Clamator Jacobinus*	F
46	Eurasian Goldan Oriole; *Oriolus Oriolus*	F
47.	Little swift; *Apus affinis*	F
48.	Asia palm Swift; *Cypiurus balasiensis*	F
49.	Streak throated swallow; *Hirundo Fluviola*	F
50.	Duskey- Martin; *Hirundo Concolor*	F
51.	Peregrine (Barbary) Falcon; *Peregrinus babylonius*	F
52.	Shikra; *Accipiter badius*	F

F = Forest; Sb = Shore bird

Table 2: Waterfowl populations (resident and migratory) recorded in the Mansagar reservoir and Check dam during May 2005 – March 2007.

S.No.	*Common & Scientific Name*	*Niche*
1.	Black Winged Stilt : *Himantopus himantopus*	Sb
2.	Red Wattled Lapwing : *Vanellus indicus*	Sb
3.	Common Coot : *Fulica atra*	D
4.	Common Moorhen: *Gallinula chloropus**	D
5.	Grey Heron: *Ardea cinerea**	Sb
6.	Little Egret: *Egretta garzetta*	Sb, W
7.	Intermediate Egret: *Mesophoyx intermedia*	Sb, W

contd...

S.No.	*Common & Scientific Name*	*Niche*
8.	Cattle Egret: *Bubulcus ibis*	Sb
9.	Little Cormorant: *Phalacrocorax niger**	D
10.	Great Cormorant: *Phalacrocorax carbo**	D
11.	Indian Pond Heron: *Ardeola grayii*	Sb
12.	Common Sandpiper: *Actitis hypoleucos**	Sb
13.	Wood Sandpiper: *Tringa glareola**	Sb
14.	Little Grebe: *Tachybaptus rufficollis*	D
15.	Little Stint: *Calidris minuta**	Sb
16.	Ruff: *Philomachus pugnax**	Sb,
17.	Little Ringed Plover: *Charadrius dubius**	Sb
18.	Northern Shoveller: *Anas clypeata**	D
19.	White-Breasted Waterhen: *Amaurornis phoenicurus*	Sb
20.	Marsh Sandpiper: *Tringa stagnatilis**	Sb
21.	Common Teal: *Anas crecca**	D
22.	Spot – Billed Duck: *Anas formosa**	D
23.	Black Crown Night Heron: *Nycticorax nycticorax**	Sb
24.	Purple Heron: *Ardea purpurea**	W
25.	Painted Stork: *Mycteria leucocephala**	W
26.	Lesser Flamingo: *Phoenicopterus minor**	W
27.	Green Sandpiper:*Tringa ochropus**	Sb
28.	Common Snipe: *Gallinago nemoricola**	Sb
29.	Common Tern: *Sterna birundo*	Sb
30.	White-Browed Wagtail:*Motacilla madaraspatensis*	Sb
31.	White Wagtail: *Motacilla cinerea*	Sb
32.	Yellow Wagtail: *Motacilla flava*	Sb
33.	Citrine Wagtail: *Motacilla citreola*	Sb
34.	Kentish Plover: *Charadrius alexandrinus**	Sb
35.	Red Shank: *Tringa tetanus*	Sb, W
36.	Black Tailed Godwit: *Limosa limosa**	W
37.	Pied Avocet: *Recurvirostra avosetta**	W
38.	Common Greenshank: *Tringa stagnatilis**	W
39.	Great White Pelican: *Pelecanus onocrotalus**	D
40.	Darter *Plotus anhinga*	D

* = Migratory; Sb = Shore bird; W = Wader; D= Diver;

Sacred groves playing greater role in conserving floral and faunal diversity Mansagar Environ are; Shree Vir Hanuman temple, Sita Ram temple and Annapurna devi temple. The presence of priests and devotees deter people from causing harm to flora and fauna. Both feed (grains) and water are provided to birds having a positive effect on their counts. Plum Headed Parakeet, an endangered species, visits regularly in large numbers at Sita Ram Temple. It is a very beautiful bird worth clicking a shot, but at a safe distance. This temple also has an endangered plant species *Commiphora wightii.* The regeneration of *Anogeissus pendula* – the key stone species of forest has also been observed. Annapurna Devi temple lying in the downstream of the lake has both swampy and marshy vegetation. It has good stand of *Ficus glomerata*, *Mitragyna parviflora* and *Phoenix sylvestris*. *Pandanus* is the important marshy species found here only. Panther sightings in the marsh are common, especially during summer. The other notable practice in the marsh are harvest of vegetation for fodder; a good example to teach nutrient recycling in per urban ecosystem.

CONCLUSIONS

The study area have numerous scope of studying, admiring, and enjoying the scenery, its wild plants and animals, as well as existing cultural manifestations (Jalmahal) found in these areas. Ecotourism aims at promoting environmental values and ethics and preserving nature in its uninterrupted form. It thus benefits wildlife and nature by contributing towards ecological integrity. Thus, there is fair scope of nature based ecotourism that needs to be tapped for economic benefits as well as for healthy conservation and preservation of nature.

ACKNOWLEDGEMENTS

One of the authors (P. K. Gputa) express gratitude to Prof. K.P. Sharma, Botany Department, UOR, Jaipur for their incessant support and guidance throughout the tenure of work. We thank my Co-supervisor Dr.Subhasini Sharma, Zoology Department, UOR, Jaipur for identification of faunal diversity, Mr.Roop singh, Curator, Herbarium Section for plant identification and Rajasthan state Forest Department, Jaipur for their cooperation in conducting the present study.

REFERENCES

1. Ali, Salim.1974. Bharat Ke pakshi Haryana Hindi Granth Academy, Chandigarh,pp.335.
2. Apate, S. A, Kumbhar, S. N. Terdalkar. S.S. and Kulkarni, A.S. 2005. Ecotourism potential of Ratnagiri Coast with special reference to Bhatye estuary. Nat. Environ. Pollut. Tech.4 (3): 363-365.
3. Bhandari,A.J.1988.Regional influence on small scale vegetational heterogeneity within grassland in the Serengeti National Park, Tanzania.Vegetatio 74:7-10.
4. Boo, E. 1992.Tourism and the Environment: Pitfalls and Liabilities of Ecotourism Development. WTO News, 9 October, pp. 2-4.
5. Boo, E. 1990. Eco-tourism; The Potential and Pitfalls (2 volumes).World Wildlife Fund, Washington, D.C.
6. Botanical Survey of India.1983.Flora and vegetation of India- an outline. BSI, Howrah, 24pp.
7. Buckley, R. 2000. Tourism in the most fragile environmens.Tourism Recreation Research 25:31-40
8. Cruickshank, A. D. 1960. A Pocket Guide to birds.Wasington Square Press, Inc. New York, pp .216.
9. Gaul, D. 2003. Environmental Impacts of Ecotourism A review of literature. FAO, Rome.
10. Kazmierczak. K. and Perlo, B.V.2000. A Field Guide to the Birds of India, Sri Lanka, Pakistan, Nepal, Bhutan, Bangladesh and he Maldives.OM Book Service, New Delhi.pp.351.
11. Kumar, A., Sati, J.P., Tak, P.C. and Alfred J.R.B.2005. Handbook on Indian Wetland Birds and Their Conservation. Zoological Survey of India, Ministry of Environment and Forest, Govt. of India, Kolkata, pp. 468.
12. Panndi, B.R., Patel, D. and Pandya, U.2005. Biodiversity measurement of Jessore sloth beer Sanctuary. Nat.Environ. Pollut.tech.4 (1): 113-118.
13. Mihalic, T.2000. Environmental management of a tourist destination's factor of tourism competitiveness. Tourism Management 21: 65-78.

Floral Diversity and their Conservation (2013), Editors: D.R. Khanna et al.
Published by Biotech Books.
ISBN: 978-81-7622-286-0 Pages: 211-225

22

Phyto-sociology, Primary Production and Nutrient Retention in Woody and Herbaceous Vegetation on the Nahargargh hill Forest nearby Mansagar Reservoir at Jaipur

P.K. Sharma and K.P. Sharma
Department of Botany, University of Rajasthan, Jaipur, India

Present study was made at three different sites (foot hill, middle hill and top hill) of the site-1 and site-2 in the rainy. 40 species of grasses and forbs were recoded in the herbaceous vegetation in forest on Nahargargh hills (Aravalli hills) nearby Mansagar Lake. *Achyranthes aspera, Brachiaria ramosa, Blainvillea acmella, Catharanthus pusila, Cassia absus, Commelina undulata, Hibiscus lobatus, Leucas aspera, Peristrophe bicalyculata* and *Triumfetta rhomboidea* were the dominant species contributing maximum to both IVI and standing crops of vegetation in the rainy season. The forest had 14 woody species *Anogeissus* being the keystone species on the middle and top hill sites of the forest while *Prosopis juliflora* on the foot hill. Litter fall contributes nutrients to soil in winter but practice of collecting *Anogeissus* litter in winter as fodder by villagers is detrimental to forest ecosystem. It also affects food security of wild herbivores in the forest, for them leaf litter constitutes major supply of food. Biomass of herbaceous vegetation at Site-1 was higher than that at Site-2 in the rainy season. Tissue concentration of nutrients in grasses and forbs were higher in the rainy season. Further, TKN and TP concentrations were higher in grasses in comparison to forbs whereas Na in the litter. K and Ca concentrations were however, similar in them. Their concentration however, differed little in post rainy season. Litter fall contributed maximum minerals at foot hill. Their standing crops decreased almost 75-80 percent at the middle hill and top hill sites. *Anogeissus pendula* litter contributed maximum to the nutrient pool (50–60 %) at the middle and top hills whereas *Prosopis juliflora* and mixed woody litter had maximum inputs at the foot hill. Maximum litter biomass was recorded at foot hill site which decreased markedly at both middle (73.4 %) and top hill (72.2 %) site. The forest floor vegetation plays an important role in nutrient cycling, habitat conservation and regeneration of tree and shrubs. The herbaceous species locks nutrients during their active growing period (rainy season) which otherwise would have been lost in runoff. Thus, they conserve nutrients in forest ecosystem.

Key words: Nahargarh hill forest, phyto-sociology, biomass, woody & herbaceous vegetation.

INTRODUCTION

The forest floor vegetation plays an important role in nutrient cycling, habitat conservation and regeneration of tree and shrubs. The herbaceous floor vegetation has been reported to show high nutrient content and rapid turnover rates as influenced by climatic condition (Spain 1984) and vegetation characteristics (Vogt & Vogt 1986). There are only few reports on the organic matter production of the ground vegetation in the forest ecosystems (Pradhan & Dash 1984; Singh & Singh 1980 and Vyas &Vyas 1978), with little information on phytosociology. I therefore, made detailed ecological studies of forest on Aravallis around lake and important findings are reported here. Due to marked differences in the topographic conditions, the vegetation composition was studied on the foot hill, middle hill and top hill.

STUDY AREA

Jaipur is located on 26° 55' north latitude and 75° 49' east longitude. The founder, Sawai Jai Singh II constructed Jaipur in the year 1727 on a sandy flat plain in the corner where the two anticlinal ranges of the Aravallis, the Galta-Jhalana on the east and the Nahargarh on the north, converge. These hill ranges of the Pre-Cambrian age were rejuvenated during the Cretaceous period. Their average height is of 500 meters and the highest peak of Jhalana range is 589 meters and in the Nahargarh 587 meters. Their flat tops of hard and resistant quartize provided suitable sites for fortifications. Climatically, Jaipur is in the semi-arid region of India. The mean temperature of Jaipur is 36^{0}C, attaining maximum in June (45-48^{0}C) and minimum in January reaches to subfreezing temperature in night. The normal rainfall of the Jaipur is 60 cm, nearly 90% of which occurs during the summer monsoon *i.e.* from June to September while the rest comes from the winter cyclones. August is the rainiest month, when the relative humidity rises to a maximum of 84%.

MATERIAL AND METHOD

Phytosociological studies of the herbaceous communities of Aravalli hills were carried out on the 6 study sites, three each located in the site -1 and site-2 in he Rainy and post rainy season (2006) and post rainy season (2007). Phytosociological Study of the woody Perennials (March 2006) were studied on the 6 sites, three each located in the site -1 and site -2.

The vegetation composition was studies on he foot hill, middle hill and top hill. Both analytical and synthetic characters of plant communities were analyzed using quadrat method. The size of quadrat for herbaceous communities was 50 x 50 cm, while for shrubs and trees 5 x 5m. The quadrats were laid randomly in the community and the data are based on average of 10 quadrats per site.

Coefficient of similarity (Sorensen Coefficient) was calculated as follows:

$$CN = 2\,J / (a + b)$$

Where, J is the number species common to both habitats while a and b are the total number species in habitat a and b respectively.

Dry matter production of the herbaceous vegetation on hills was determined by harvest method. Five quadrats (50 cm × 50 cm) were laid randomly on foot, middle and top hill sites at two occasions in the year 2006 (rainy and post - rainy season) while only in the post-rainy season in

2007. All shoots were harvested along with their roots up to a depth of 10cm depth since habitat was rocky. Roots were washed carefully in the troughs to remove adhering soil. Harvested plants were separated species wise, dried in hot air over at 60^0C for one week and weighed (shoot + root).

Litter was collected carefully in winter by laying quadrat (50 × 50 cm) randomly on the forest floor and separated species wise for the key stone species, dried in hot air oven and weighed, as described earlier.

Analysis of herbaceous vegetation and plant litter were made. The oven dried vegetation was pooled in two categories as forbs and grasses. Similarly litter was separated as leaf litter of *Anogeissus*, *Prosopis*, mixed leaf litter and woody litter. All these categories were powdered separately in a mixer which was digested in DIV and TKN, TP, Ca, Na and K therein were analyzed using standard methods

Soil samples collected (2006) from foot hill, middle and top hill sites of Jhalana hills were dried in the sun, ground and then sieved through 80 mesh sieve and than analyzed for their physico-chemical characteristics following methods described by Jackson (1967).

RESULT AND DISCUSSION

The ground flora of forest in rainy season a site-1 contains of 25 species. Species diversity at the foot and middle hill sites was similar (12 species) while it increased moderately (18 species) at top hill site. Among 25 species, following 7 species contributed maximum IVI. *Commelina undulata, Brachiaria ramosa, Achyranthes aspera, Peristrophe bicalyculata, Catharanthus pusila, Cassia absus* and *Hibiscus lobatus.*

On site-2, the ground flora of forest in the rainy season consisted of 22 species. Species diversity at foot and middle hill sites was similar (11 species) while it increased a little (14) at top hill. Among 22 species, following 8 species contributed maximum to IVI (Importance value index). *Commelina undulata, Catharanthus pusila, Sesamum indicum, Lantana indica, Euphorbia parviflora, Lucas aspera, Triumfetta rhomboidea* and *Blainvillea acmella.*

Commelina, Euphorbia and *Blainvillea* were found to be the most dominant species respectively at foot, middle and top hill sites. *Euphorbia* was replaced by *Triumfetta* at mid hill while *Blainvillea* by *Sesamum* when biomass was used for calculating relative dominance.

In the rainy season on site-1,closeness between ground flora of foot hill was relatively higher with middle hill ground flora as almost 50 percent plant species were common to them which decreased to about 33 percent in case of top hill flora. In contrast middle hill ground flora had almost equal similarity with both foot hill and top hill flora having almost 50 percent common species. It is evident that middle hill represents transition zone between foot hill & top hill. On site-2, closeness between ground flora of foot hill was relatively higher with middle hill ground flora as almost 55 percent plant species were common to them which decreased to about 33 percent in case of top hill flora . In contrast middle hill ground flora had almost equal similarity (55- 58%) with both foot hill and top hill flora, and therefore, it was a transition zone. A comparison of site-1 with site-2 revealed marked dissimilarity in the ground flora of foot and middle hills since only 41-52 percent of the species were common. In contrast, top hill flora at these sites had 64 percent common species.

Table 1: IVI of herbaceous ground flora of forest at Site 1and Site-2 in the rainy season.

Species	Site-1			Site-2		
	Foot hill	**mid hill**	**Top hill**	**Foot hill**	**mid hill**	**Top hill**
Achyranthes aspera Linn.Var. argentea Hk. f.	21.76	31.93	21.1	Ab.	11.97	Ab.
Aristida adscensionis Linn.	Ab.	Ab.	9.33	Ab.	Ab.	13.37
Blainvillea acmella (Linn.) Philipson	Ab.	Ab.	4.57	21.61	26.94	67.9
Borreria stricta (Linn.f.) Schum	15.04	Ab.	Ab.	9.99	Ab.	Ab.
Brachiaria ramosa (L.) Stapf.	24.85	32.63	15.05	9.15	8.68	8.37
Cardiospermum halicacabum L. var. microcarpum (Kunth) Blume.	Ab.	Ab.	4.63	Ab.	Ab.	11.23
Cassia absus Linn.	Ab.	Ab.	24.81	Ab.	Ab.	6.34
*Catharanthus pusila (*Murr.) G. Don	14.98	31.59	78.29	Ab.	7.39	30.88
Corchorus aestuans L.	10.88	20.17	Ab.	Ab.		Ab.
Commelina undulata R.Br.	123.25	89.76	64.62	123.91	Ab.	46.66

contd...

Cypress compress Linn.	Ab.	25.93	11.9	Ab.	Ab.	Ab.
Hibiscus lobatus (Murray) Kuntze	13.38	Ab.	Ab.	Ab.	Ab.	Ab.
Euphorbia parviflora Linn	Ab.	9.37	5.39	15.03	95.1	7.63
Indigofera astragaline D.C.	Ab.	Ab.	11.06	Ab.	Ab.	Ab.
Indigofera cordifolia Heyne ex Roth.	Ab.	Ab.	4.64	Ab.	Ab.	Ab.
Ipomoea nil (Linn) Roth.	6.77	Ab.	Ab.	Ab.	Ab.	Ab.
Ipomoea eriocarpa R.Br.	Ab.	6.83	Ab.	Ab.	14.61	13.51
Leucas aspera (Willd.) Spreng.	Ab.	10.14	7.88	28.91	32.77	8.37
Physalis minima Linn	Ab.	Ab.	Ab.	16.54	Ab.	Ab.
Peristrophe bicalyculata (Retz.) Nees.	27.13	Ab.	Ab.	Ab.	Ab.	Ab.
Pupalia lappacea (Linn.) Juss.	Ab.	Ab.	5.98	Ab.	Ab.	Ab.
Sida veronicaefolia Lam.	8.86	16.54	Ab.	15.7	Ab.	Ab.
Tephrosia strigosa (Dalz.) Sant. et Maheshw.	Ab.	Ab.	Ab.	Ab.	Ab.	6.51
Tridax procumbens Linn	Ab.	Ab.	Ab.	Ab.	7.41	7.01
Triumfetta rhomboidea Jacq.	20.04	Ab.	6.85	29.71	74.18	9.8
Vernonia cinerea (Linn.) Less.	11.15	Ab.	Ab.	9.61	13.12	Ab.
Lantana indica Linn.	Ab.	Ab.	Ab.	Ab	7.83	Ab.
Composite member	Ab.	Ab.	Ab.	19.89	Ab.	Ab.
Zornia gibbosa Span.	Ab.	Ab.	4.82	Ab.	Ab.	Ab.
Tetrapogon tenellus (Roxb.) Chiov.	Ab.	15.9	10.1	Ab.	Ab.	Ab.
Sesamum indicum Linn.	Ab.	9.47	8.7	Ab.	Ab.	62.6

Biomass of herbaceous vegetation at Site-1 was higher than that at Site-2 in the rainy season. Biomass was found maximum at the top hill that decreased moderately at foot hill but only little at middle hill .The maximum biomass at Site-2 was also recorded at middle hill and percentage reduction in biomass was higher at foot in comparison to top hill.

At Site-1, species wise *Achyranthes aspera, Commelina undulata, Catharanthus pusila, Indigofera astragaline, Ipomoea iriocarpa, Sesamum indicum* and *Brachiaria ramosa* contributed maximum (39.3-72.9%) to biomass in the rainy season.

In all 12 woody species were recorded. Their species richness was found maximum at the foot hill (8 species), which decreased a little at the middle and top hill (6 species) on site-1.

Based on IVI, following associations of plant species were characterized.

Foot hill: *Prosopis juliflora - Adhatoda vasica - Anogeissus pendula - Lantana indica*

Middle hill: *Anogeissus pendula - Euphorbia neriifolia- Azadirachta indica - Prosopis juliflora*

Top hill: *Anogeissus pendula -Euphorbia neriifolia - Grewia tenax - Prosopis juliflora.*

On site-2, In all 12 woody species were recorded. Their species richness was maximum at the foot hill (10 species) which decreased a little at the middle and top hill (6 species).

Based on IVI, following associations of plant species were characterized

Foot hill: *Prosopis juliflora- Anogeissus pendula- Adhatoda vasica- Acacia senegal - Azadirachta indica*

Middle hill: *Anogeissus pendula- Prosopis juliflora- Euphorbia neriifolia-Capparis callosa - Grewia tenax*

Top hill: *Anogeissus pendula- Grewia tenax- Euphorbia neriifolia- Acacia senegal - Prosopis juliflora*

It is evident that *Prosopis juliflora* was the most dominant species at the foot hill site whereas *Anogeissus pendula* at middle and top hill

Sorensen coefficient on site-2, closeness between woody species at foot hill was similar with middle and top hill flora as 44 percent plant species were common to them. In contrast, middle hill ground flora had more similarity with top hill having about 83 percent plant species in common which decreased to about 44 percent with foot hill flora. A comparison of site-1 with site-2 revealed higher degree of closeness between.

Table: IVI of Woody species at site-1 and site-2

Plant species	*Site-1*			*Site-2*		
	Foot hill	*Middle hill*	*Top hill*	*Foot hill*	*Middle hill*	*Top hill*
Prosopis juliflora (Sw.) DC. var. *juliflora*	111.2	6.8	13.4	119.4	35.8	7.7
Anogeissus pendula Edgew.	68.5	226.5	213.4	109.9	234.7	242.3
Adhatoda vasica Nees	82.1	5.02	Ab.	23.7	Ab.	Ab.
Lantana indica Roxb	16.2	Ab.	Ab.	5.5	Ab.	Ab.
Abutilon indicum (L.) Sweet.	3.52	Ab.	Ab.	5.5	Ab.	Ab.
Oxystelma secamone (Linn.)Karst	7.8	Ab.	Ab.	3.8	Ab.	Ab.
Azadirachta indica A. Juss.	6.8	27.4	Ab.	12.1	Ab.	5.8
Acacia senegal (L.) Willd	Ab.	Ab.	Ab.	13.1	5.6	11.8
Capparis callosa Blume	3.9	Ab.	Ab.	4.7	5.8	Ab.
Grewia tenax (Forssk.) Fiori.	Ab.	5.2	23.6	Ab.	5.6	25.5
Moringa olcifera Lamk.	Ab.	Ab.	14	Ab	Ab.	Ab.
Capparis sepiaria Linn.	Ab.	Ab.	5.6	Ab	Ab	Ab.
Calotropis procera (Alt.) R.Br	Ab	Ab	Ab	5.1	Ab	Ab.
Euphorbia neriifolia Linn.	Ab.	28.9	29.8	Ab.	12.0	12.6

Based on Sorensen coefficient, closeness between woody species a foot hill was relatively higher with middle hill flora as almost 57 percen plant species were common to them which decreased markedly with To hill as only 28 percent species were common between them. In contras middle hill ground flora had similar closeness with top and foot hill flor having 57-66 percent plant species in common. It is thus evident that middl hill represents transition zone between foot hill & top hill on site-1.Base on woody floras of their respective sites. Further, woody flora of middl hills had higher similarly with that of top hills of these sites

Soil pH differing little at foot, middle and top hill of Site 1 and 2 was i alkaline range (7.82 - 8.22) in the rainy season. EC, TKN, TP, exchangeab phosphorus, Na, K and Ca had differed little in the rainy season at th two sites. Chlorides and Sulphate concentration was higher on site-2 i comparison of site- 1. Organic carbon had higher concentration on site-1 i comparison to site-2.

Table: Physico-chemical characteristics of forest soil in the rainy season.

Site		*pH*	*EC*	*Cl*	*SO_4*	*Org. C. (%)*	*TKN*	*TP*	*IP*	*Na*	*K*	*Ca*
S1	**Foot**	8.06	0.48	3.7	2800	1.14	13.1	35.2	5.43	41.5	512.5	56
	Mid	8.06	0.34	2	3600	1.655	22.5	34	7.81	49.5	412.5	67
	Top	8.04	0.34	5.7	2400	1.655	27.4	40.2	4.9	56.5	525	70.5
S2	**Foot**	7.82	0.38	8	4000	0.206	34.5	27.1	0.83	54	487.5	65
	Mid	7.96	0.24	5	4000	1.189	24.9	31.2	4.8	50	375	68
	Top	8.22	0.27	9.5	5200	0.465	27.2	26.5	2.9	50	487.5	68.5

Except for pH, EC (μ S) and organic carbon, all values are in mg/100g:

Tissue concentration of nutrients in grasses and forbs were higher in the rainy season. Further, TKN and TP concentrations were higher in grasses in comparison to forbs whereas Na in the latter. K and Ca concentrations were however, similar in them. Their concentrations however, differed little in post rainy season.TKN standing crop at Site-1 was higher (2.08-4.89 g m^2) in comparison to Site-2 (0.6-2.81 g m^2). Further, its maximum standing crops were recorded in the rainy season at both sites. The maximum standing crops of TKN on site-1 were recorded at the middle hill (2.43-4.89 g m^2) and minimum at the foot hill (2.08-2.84 g m^2).

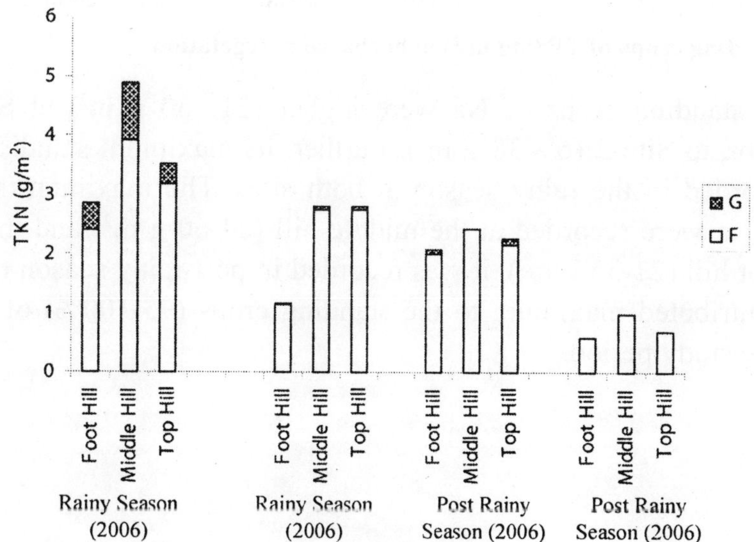

Fig. 1: Standing crops of TKN (g m^{-2}) in herbaceous vegetation

The standing crops of phosphorus were higher (0.35-0.69 g m^2) at Site-1 in comparison to Site-2 (0.09 - 0.38 g m^2). Further, its maximum values were recorded in the rainy season at both sites. The maximum standing crops of phosphorus on site-1 were recorded at middle hill (0.69 g m^2) and minimum at foot hill (0.39 g m^2) in the rainy season. In post rainy season, the maximum standing crops of phosphorus were recorded at the top hill (0.59 g m^2) and minimum at the foot hill (0.35 g m^2).

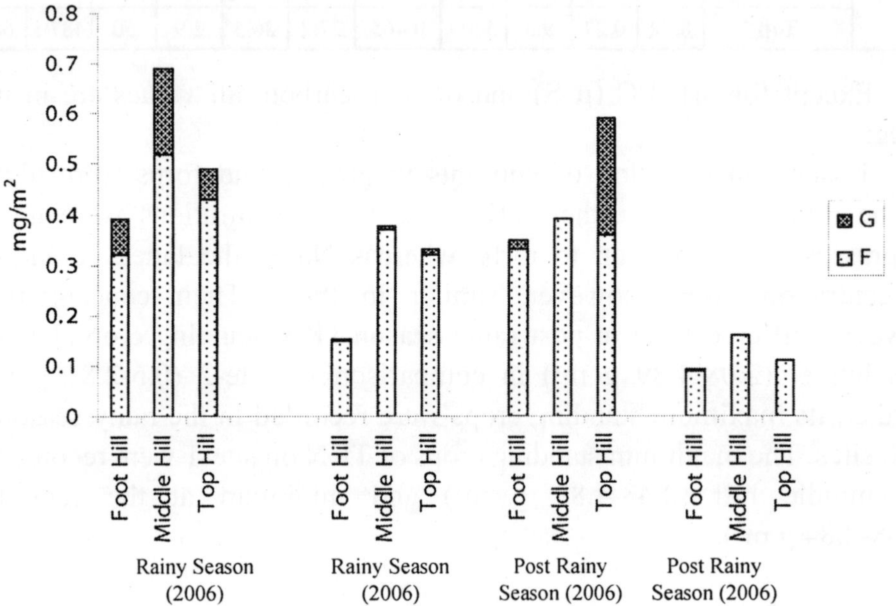

Fig. 2: Standing crops of TP (mg m^{-2}) in herbaceous vegetation

The standing crops of Na were higher (21- 60 g m^2) at Site-1 in comparison to Site-2 (6 - 38 g m^2). Further, its maximum standing crops were recorded in the rainy season at both sites. The maximum standing crops of Na were recorded at the middle hill (24-60 g m^2) and minimum at the foot hill (21-35 g m^2). it was recorded in post rainy season (30.7%). Forbs contributed maximum to the standing crops (95-100%) of sodium during the study period.

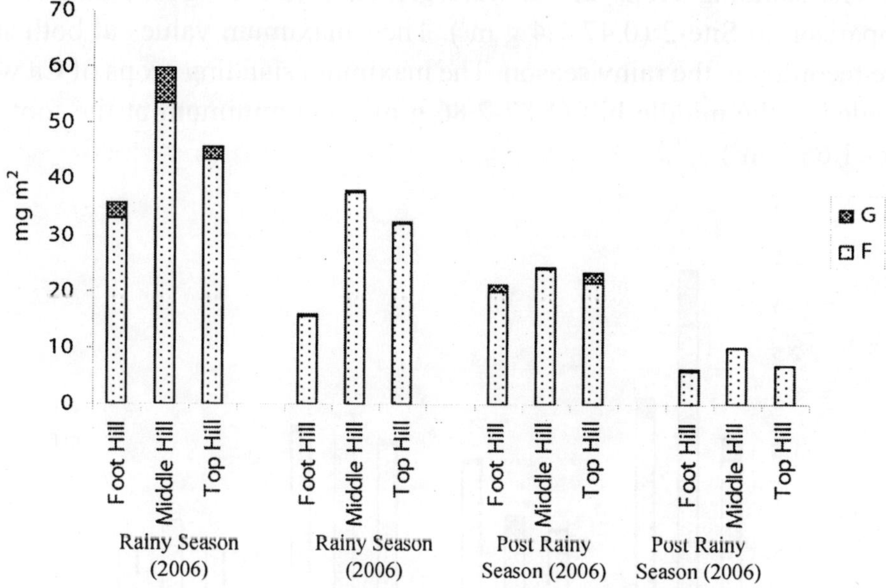

Fig. 3: Standing crops of Sodium (mg m^{-2}) in herbaceous vegetation.

The standing crops of K was higher (4-9 g m^2) at Site-1 in comparison to Site-2 (1.7 -5.0 g m^2). Further, its peak values were recorded in the rainy season at both sites. On site-1,the maximum standing crops of K were recorded at middle hill (4.14 - 9.04 g m^2) and minimum at foot hill (3.59 - 5. 28gm^2) .

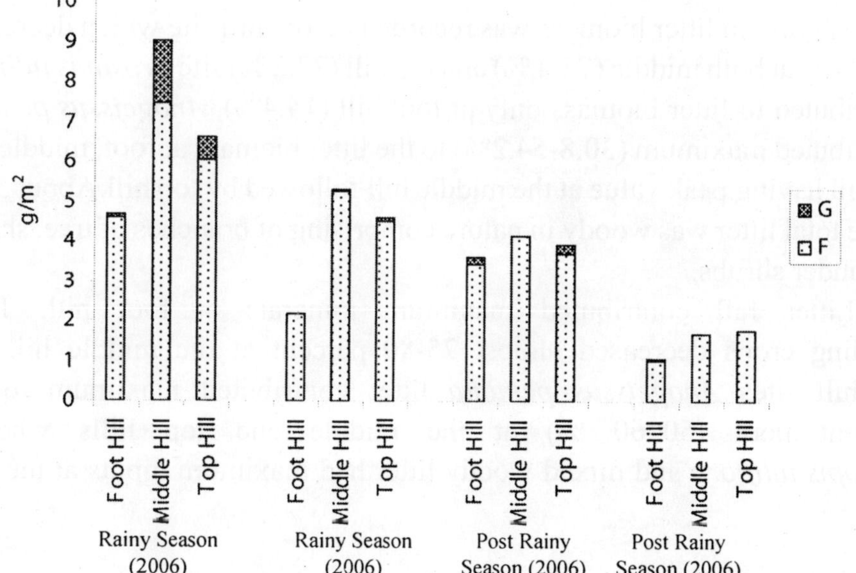

Fig. 4: Standing crops of Potassium (gm^{-2}) in herbaceous vegetation.

The standing crops of Ca were greater (1.6-2.86 g m^2) at Site-1 in comparison to Site-2 (0.47-1.4 g m^2). Their maximum values at both sites were recorded in the rainy season. The maximum standing crops of Ca were recorded at the middle hill (1.87-2.86 g m^2) and minimum at the foot hill (1.6 - 1.65 g m^2)

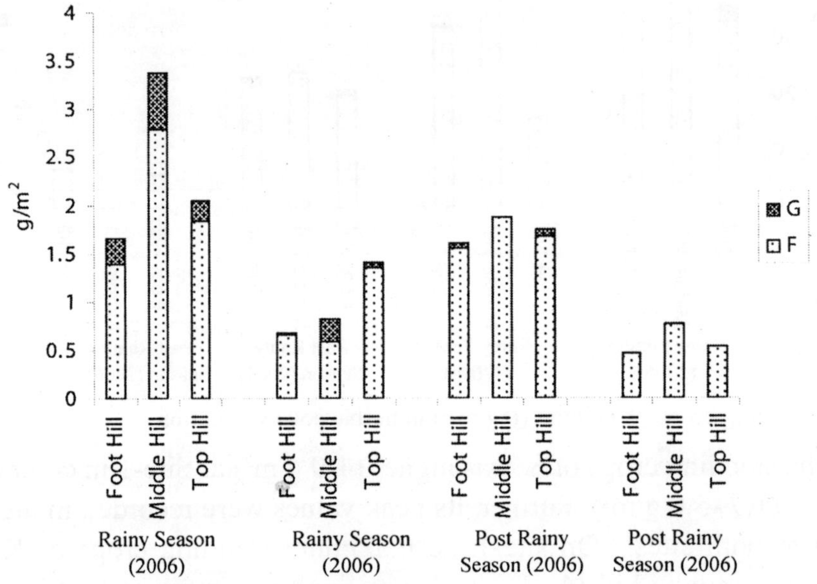

Fig.5: Standing crops of Calcium (gm^{-2}) herbaceous vegetation

Maximum litter biomass was recorded at foot hill site which decreased markedly at both middle (73.4 %) and top hill (72.2 %) site *Prosopis juliflora* contributed to litter biomass only at foot hill (14.4%). *Anogeissus pendula* contribuied maximum (30.8-54.2%) to the litter biomass at foot, middle and top hill having peak value at the middle hill followed by top hill.About 50 % of the total litter was woody in nature comprising of branches of tree, shrubs and under shrubs.

Litter fall contributed maximum minerals at foot hill. Their standing crops decreased almost 75-80 percent at the middle hill and top hill sites *Anogeissus pendula* litter contributed maximum to the nutrient pool (50–60 %) at the middle and top hills whereas *Prosopis juliflora* and mixed woody litter had maximum inputs at the foot hill.

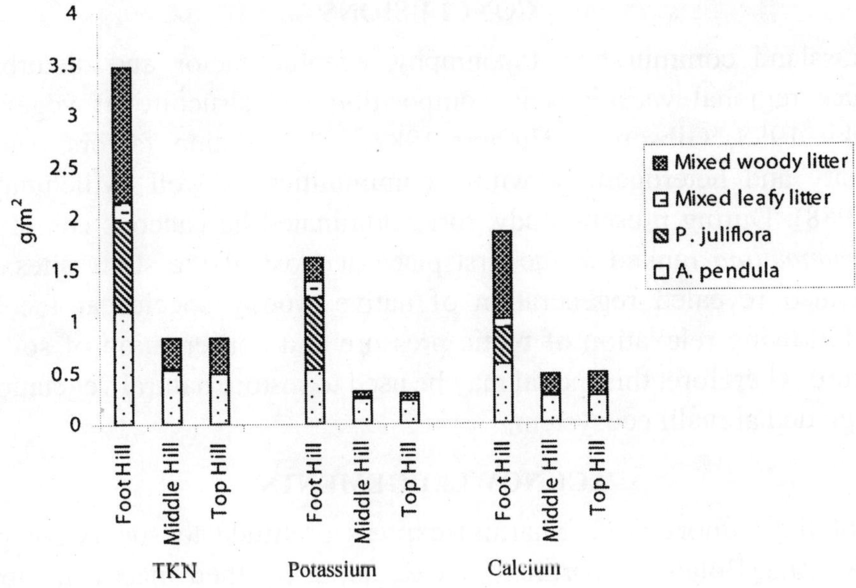

Fig.6: Standing crops of TKN, K and Ca (gm^{-2}) in litter

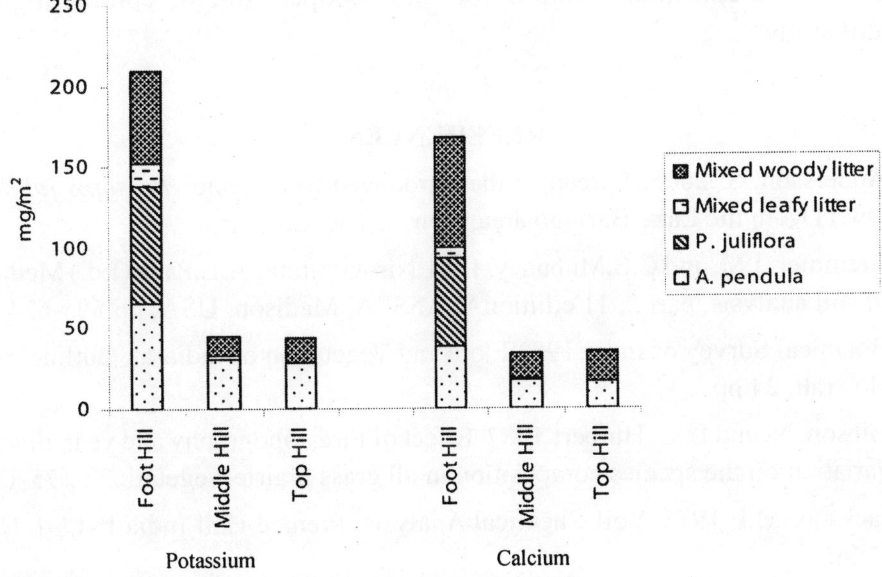

Fig.7: Standing crops of TP and Sodium (gm^{-2}) in litter

CONCLUSIONS

In grassland communities, topography, edaphic factor and disturbance produce regional variations in composition and structure of vegetation (Belsky 1988, Gibson & Hulbert 1987). These same factors increase diversity and heterogeneity within communities as well (Milchunus *et al.*. 1988). During present study, forbs dominated herbaceous community and *Commelina* ranked at the first place at most of the study sites. The study also revealed regeneration of native woody species at the both sites following relaxation of biotic pressure and conservation of soil and moisture. Therefore, this modal may be used to restore natural vegetation in he degraded aravalli ecosystem.

ACKNOWLEDGEMENTS

One of the authors (P.K. Sharma) express gratitude to supervisor prof. .K.P.Sharma, Botany Department, UOR, Jaipur for their incessant support and guidance throughout the tenure of work. Thanks are expressed to the head, Botany Department, UOR, Jaipur for laboratory facilities, Mr.Roop singh, Curator, Herbarium Section for plant identification and Rajasthan state Forest Department, Jaipur for their cooperation in conducting the present study.

REFERENCES

1. Andersson, S. 2005, Spread of the introduced tree species *Prosopis juliflora* (sw.) DC in the Lake Baringo area, Kenya. Thesis.
2. Bremmer, J.M. and C.S.Mubaney, 1981.Nirogen-total.A.L.Page (Ed.) Methods of soil analysis, part 2, 11 ediitionASASSSA, Madison, USA, pp.595-624.
3. Botanical Survey of India.1983.Flora and Vegetation of India-an Outline. BSI, Howrah, 24 pp.
4. Gibson, N. and L. C. Hulbert.1987. Effect of Fire, topography and year climatic variation on the species composition in all grass prairie.Vegetatio72:175-185.
5. Jackson, M.L.1973. Soil Chemical Analysis. Prenice Hall India Pvt.Ld. New Delhi.
6. Joshi, S. K., D. P. Pati and N. Behera. 1990. Primary production of herbaceous layer in a tropical deciduous forest in Orissa, India. Tropical Ecology 31:73-83.

7. Joshi, S. P., A. Raizada and M.M.Srivastava.1988. Net primary productivity of high altitude grassland in Garhwal, Himalaya. Tropical Ecology 29:15-20.
8. Belsky, A. J. 1988. Regional influence on small scale vegetational heterogeneity within grasslands in he Serengeti National Park, anzania. Vegetatio 74: 7-10.
9. Khairnar, D. N. 2005. Biodiversity of rare and endangered plan of East Nasik region (North Sahyadri), Maharashra.Na.Environ.Pollu.Tech.4 (4):585-587.
10. Odiwe, A. I. and Muoghalu, J. I. 2003.Litterfall dynamics and forest floor litter as influenced by fire in a secondary Lowland rain forest in Nigeria.rop. Ecol.44:241-248.
11. Pandit, B. R., Patel, D. and Pandya, U. 2005 Biodiversity Measurement of Jessore sloth beer Sanctuary. Nat. Environ. Pollu. Tech. 4(1):113-118.
12. Pradhan, G. B. and M.C. Dash. 1984. Seasonal variations in plant biomass and net primary productivity of a savanna type hill ecosystem of Samalpur, India. Tropical Ecology 25:172-178.
13. Spain, K. P. and R. P. Singh. 1980. Biomass and net production of herbaceous layer in tropical dry deciduous forest a Varanasi. Tropical Ecology 21:47-58.

7. Joshi, S. P., A. Kaushal and M.M. Srivastava. 1988. Net primary production of high altitude grassland in Garhwal Himalaya. Tropical Ecology 29: 1–[illegible].

8. Belsky, A. J. 1986. Keystone influences on small-scale vegetational heterogeneity within grasslands in the [illegible] National Park, [illegible]. Vegetatio [illegible]

9. Khumbar, [illegible] 2005. Biodiversity of rare and endangered plants of East Nasik region [illegible]. [illegible] Environment and Pollution Technology [illegible]

10. Calvo, A. [illegible] 2003. Litterfall dynamics and forest floor litter as influenced by fire in a secondary lowland rain forest in Nicaragua. [illegible] Ecol. 44: [illegible]

11. Pandit, B. R., Patel, D. and Pandya, L. 2003. Biodiversity measurement of [illegible] Environment Pollution Technology [illegible]

12. Pradhan, C. B. and M.C. Dash. 1984. Seasonal variations in plant biomass and net primary productivity of a savanna type hill ecosystem of Sambalpur, India. Tropical Ecology 25: [illegible]

13. [illegible], P. and K. P. Singh. 1986. Biomass and net production of herbaceous layer in tropical deciduous forest at Varanasi. Tropical Ecology [illegible]

Floral Diversity and their Conservation (2013), Editors: D.R. Khanna et al.
Published by Biotech Books.
ISBN: 978-81-7622-286-0 Pages: 227-232

23

Profile of Lead and Mercury in Herbal Drugs from Singrauli District

*ANSHU R-ANI PATEL AND KANTISHREE DE
(M.SC., PH.D., P.D.F (C.S.I.R), F.B.S.)
Department of Post Graduate Studies and Research in Biological Sciences,
Rani Durgavati, Vishwavidyalaya, Jabalpur

Millions of Indians use herbal drugs regularly, as spices, home remedies, health foods as well as self medication. Three herbal drugs which is most popular routinely used (*Mentha arvensis, Terminalia arjuna* and *Asparagus racemosus*) were selected for this study. Heavy metals like lead and mercury were analysed in these herbal drugs in singrauli district from two different places *viz.* polluted and non polluted by the atomic absorption spectrometery. The concentration of lead was found to be 19.57 ppm in *Mentha*, 18.87 ppm in *Asparagus,* and 16.93 ppm in *Terminalia.* Likewise the concentration of mercury was 6.85 ppm in *Mentha,* 6.29 ppm in *Asparagus,* and 5.20 ppm in *Terminalia.* was observed.

Key words: lead, mercury, herbal plants, soil, Singrauli.

INTRODUCTION

Singrauli consists of north east area of Madhya Pradesh and southern part of sonebhadra district of Uttar Pradesh. It is 50th district of Madhya Pradesh. It is emerging as energy hub of India, due to availability of coal and water.

District singrauli is located between longitude 82^0 30' to 82^0 50'E and latitude 24^0 00', 24^0 20'N. Total area of it is 5672 sq.km. Vindhyanagar, Amlori, Nigahi, Jayant,Dudhichua, and jhingurada of the district are included in critically polluted area.

Millions of Indians use herbal drugs regularly, as spices, home remedies, health foods (Gautam *et al.*,2003). Heavy metals are kept under environmental pollutant category due to their toxic effects in plants, human and food. Some of the heavy metals *i.e.* Arsenic (As), Cadmium (Cd), Mercury (Hg), Lead (Pb), are cumulative poison (Gawel *et al.,* 1996). Lead (Pb) is one of the most dangerous, nonessential, bio-accumulating and toxic heavy metals. It has become the most important metal pollutant of the environment (Landrigan *et al.*,2000). They induce various toxic effects in humans at low doses. The typical symptoms of lead poisoning are colic, anemia, headache, convulsions and chronic nephritis of the kidneys, brain damage and central nervous system disorders (Salt *et al.*,1998). In the same way mercury is also considered as highly toxic contaminant. Industrial wastes and sewage water from the chloralkali industry are a major source of mercury pollutions. Mercury symptoms are fatigue, loss of memory and concentration, tremors, constriction of visual field, cortical blindness etc. (Jarup, 2003).

Terminalia arjuna (Arjuna) is a medicinal plant of the genus Terminalia, widely used by ayurvedic physicians for its curative properties in organic/ functional heart problems including angina, hypertention and deposits in arteries. It is very useful in the treatment of any sort of pain due to falls, ecchymosis, spermatorrhoea and sexually transmitted diseases such as gonorrhea (Paarekh 2010).

Asparagus racemosus, (Shatavari) plant is used traditionally for treatment of many diseases. Tuber,leaves,and fruits are used in gonorrhea,p iles,diabetes,and headache,also for increase lactation is a general tonic and also a female reproductive tonic. Shatavari is used as the main Ayurvedic rejuvenative tonic for females. Shatavari roots used as a drug acting on all tissues as a powerful anabolic. It is good for eyes, muscles, reproductive organs, increase milk secretion and helps to regain vigour and vitality (Chopra *et al.*,1986).

Mentha arvensis (Pudina) also known as Mint. Mint was originally used as a medicinal herb to treat contraceptive, carminative, antispasmodic,anti peptic ulcer agent, and has been given to treat indigestion, skin diseases, coughs and colds in folk medicine and it is commonly used in the forms of tea as a home remedy to help alleviate stomach pain (Londonkar and Poddar).

MATERIALS AND METHODS

Herbal drugs namely *Terminalia, Asparagus*, and *Mentha* were selected from two different places of Singrauli *viz.*

1) Vindhyanagar (polluted area)
2) Vindhyanagar (non polluted area)

Plant samples were digested before analysis to reduce organic matter interference and allowed the conversion of the metal into a form that can be analyzed by the atomic absorption spectrophotometer. The digested samples were cooled and diluted with distilled water and filtered through Whatman No.1 filter paper.Find Volume was made up to 50ml. After calibration the content of lead and mercury in samples were determined (Lindsay and Norvell,1978).

RESULTS AND DISCUSSION

Present study was targetted on comparative evaluation of lead (Pb) and mercury (Hg) contents in herbal drugs at two different places of Singrauli (polluted and non polluted). Three herbal drugs were taken into consideration to analyze concentration lead and mercury *viz.*

Terminalia arjuna
Asparagus racemosus,
Mentha arvensis

Table : Lead (Pb) and Mercury (Hg) concentration in three plants sample in (ppm).

Sr. No.	Places	Pb concentration			Hg concentration		
		Terminalia arjuna	***Asparagus racemosus,***	***Mentha arvensis***	***Terminalia arjuna***	***Asparagus racemosus***	***Menta arvensis***
1	Vindhyanagar (polluted area)	16.93	18.87	19.57	5.20	6.29	6.85
2	Vindhyanagar (non polluted area)	10.56	13.60	14.60	2.22	3.18	2.22

The results indicate that the highest concentration in lead was found to be in *M. arvensis* (19.57) and mercury concentration was *M. arvensis* (6.85) in polluted area.

Present study indicates that due to presence of the high heavy metal content in polluted area the metal accumulation in medicinal plants grown in the vicinity represents a potential risk for public health. There is an urgent need for studies to analyses the frequency and potential risk of heavy metal poisoning from herbal drugs and medicinal plants for culturally appropriate education to inform the public of the potential for toxicity associated with these products.

ACKNOWLEDGEMENT

The authors are thankful to the Head of the Department of Biological Science, Rani Durgavati Vishwavidalaya, Jabalpur (M.P.) for providing help in various ways for carrying out the present work. Our sincere thanks to Dr. Devendra K. Patel, Scientist of Analytical Chemistry Section, IITR Lucknow, for his kind support and providing laboratory facilities.

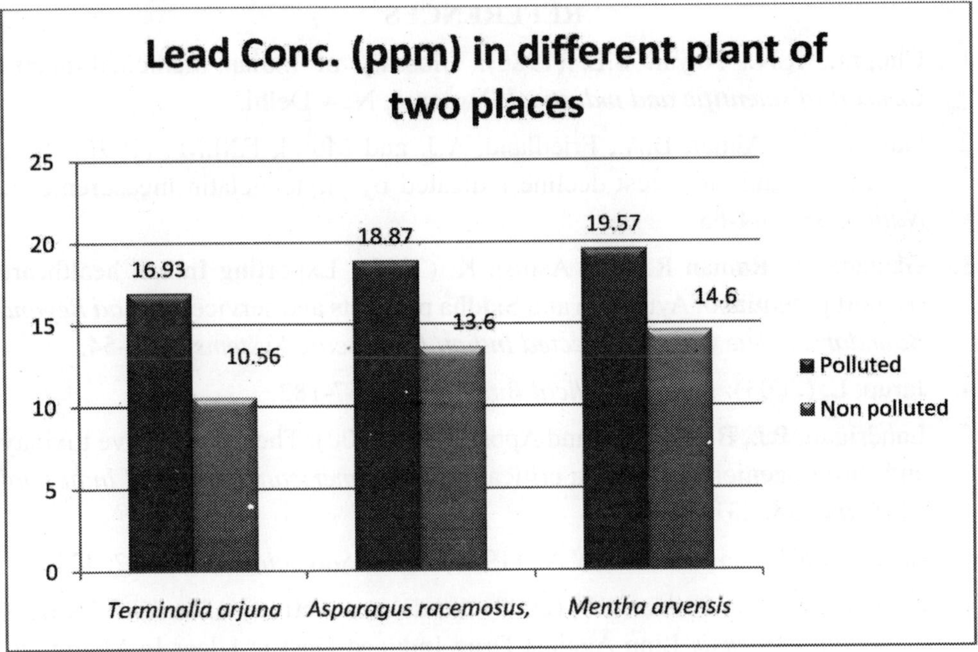

Figure 1:

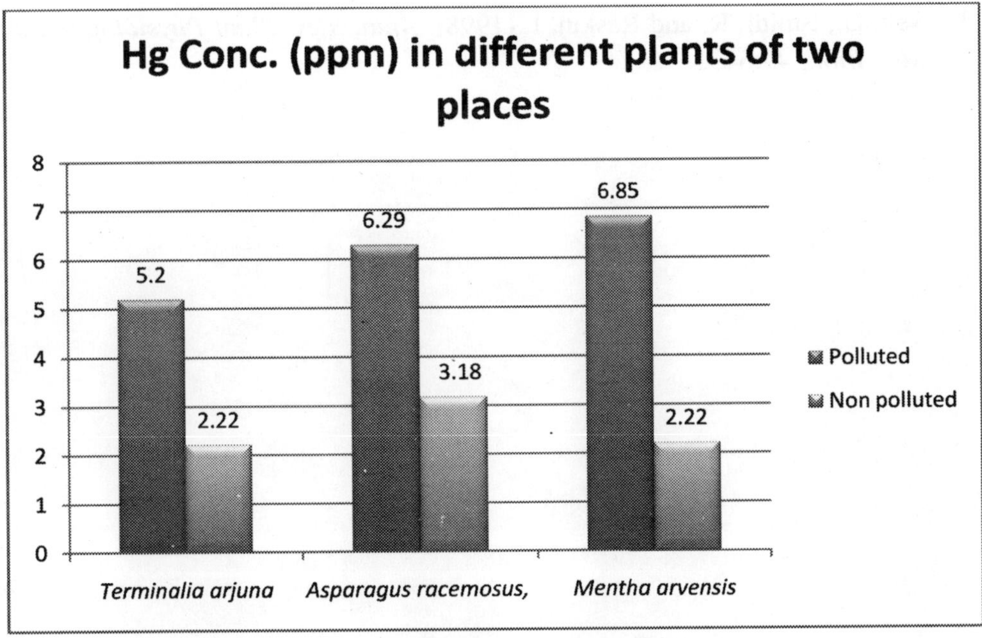

Figure 2:

REFERENCES

1. Chopra., R.N., Nayar. S.L. (1986): Glossary of Indian medicinal plants. *Council of scientific and industrial Research*, New Delhi.
2. Gawel, J.E., Ahner, B.A., Friedland, A.J. and Morel, F.M.M., (1996): Role of heavy metals in forest decline indicated by phytochelatin measurements. *Nature*. 381: 64-65.
3. Gautam, V., Raman R.M.V., Ashish K. (2003): Exporting Indian healthcare (Export potential of Ayurveda and Siddha products and services). *Road Beyond Boundaries (The Case of Selected Indian Healthcare Systems)*. 14–54.
4. Jarup, L. (2003): *British Medical Bulletin* 68 :167-182.
5. Landrigan, P.J., Boffetta, P., and Apostoli, P. (2000): The reproductive toxicity and carcinogenicity of lead: a critical review. *American Journal of Industrial Medicine*. 38: 231–243
6. Lindsay, W.L., and Norvel, W.A., (1978): *Proc. Soil science Am., 42*: 421.
7. Londonkar, R.L., Poddar, P.V. (2009): Studies on Activity of Various Extracts of *Mentha Arvensis* Linn Against Drug Induced Gastric Ulcer In Mammals. *World J Gastrointest Oncol.;* 15(1):82-88.
8. Paarakh, P.M.(2010) Terminalia arjuna (Roxb.) *International Journal of Pharmacology* 6:5 (515-534)
9. Salt, D., Smith, R. and Raskin, I. (1998): *Annu. Rev. Plant Physiology Plant Mol. Biol., 49*:643.

Floral Diversity and their Conservation (2013), Editors: D.R. Khanna et al.
Published by Biotech Books.
ISBN: 978-81-7622-286-0 Pages: 233-238

24

Depletion of Plant Biodiversity in Marathwada Region of Maharashtra State due to Bio-prospecting of Medicinal Plants

R.G. Chillawar and O.S. Rathor*
Dept. of Botany Yeshwant Mahavidalaya Nanded (M.S.) India
*Dept .of Botany Science College Nanded (M.S.) India-431605.

The rich flora of Maharashtra state includes the border region of Southern part which is known as Marathwada. The flora of Marathwada is known to have about 1700 plant species, out of which at least 300 plants are with known traditional medicinal properties. Due to sudden boom in Bio-prospecting activity about medicinal plants, a number of important medicinal plants are suddenly disappearing from the region. The present communication throws a light on these sudden losses of plant diversity and suggestive methods of conservation of these plants.

Key words: Plant bio-diversity, Bio-prospecting potential, Conservation of Medicinal plants.

INTRODUCTION

Ever since the WHO'S recognition to the traditional medicine and traditional healing systems, there is a great boom throughout the world about the traditional medicinal plants.

The medicinal plants are now traced from the point of view of bio-prospecting potential. Bioprospecting is a process through which value addition can be made to the biodiversity of planet earth. It can also be defined as the search for valuable compounds and genetic material from plants, animals and microorganisms of different ecosystems for the benefit of the society (Strenlof 2000). Although bioprospecting in the past has focused on useful organic chemicals, the search is now turning to nucleotide applications.

Bioprospecting of plants entails the search for economically valuable biochemical resources from the biological wealth of country. Such initiatives hold the promise of new medicines and biodegradable pesticides that can be a source of income for developing countries thus, can enhance the conservation value of Indian medicinal plants and also forcing incentives to conserve this bio-diversity.

Bioprospecting is involved in the exploration of currently unidentified or identified species of living organisms for the development of pharmaceuticals, identified botanical medicines, crop protection agents, cosmetics , horticultural and agricultural seeds, environmental monitoring, manufacturing and construction. The bioprospecting of plants is presently focused on medicinal plants which has following basic main areas

i) Identifying bioactive constituents
ii) Producing plant based pesticides and therapeutic compounds
iii) Studying the efficacy and safety of plant based composite medicines and
iv) Developing programmess to increases the value of raw plant materials by genetic methods.

The Marathwada region of Maharashtra state is having a rich bio-diversity of Angiospermic plants with about 1700 plant species, out of which nearly 20 % are known to have medicinal properties (Naik-1998) knowing this fact the people are investigating the localities of these useful medicinal plants and simply removing them for further investigations on bioprospecting potential, it is causing a depletion of plant bio-diversity of the region which is very much harmful.

The present paper describes the list of the medicinal plants of the area and enlists the plants which are lost due to this activity.

MATERIAL AND METHODS

Periodical surveys of the region were made and the plant flora was critically noted from time to time and causes of losses of plant were investigated which revealed that a number of plants are being over exploited without adopting any methods of conservation, because these plants are known to have a great bioprospecting potential. The herbarium specimens and all such plants are deposited in herbarium of Science College, Nanded.

RESULT AND DISCUSSION

Table No.1 enlists some of the medicinal plants of Marathwada region which are having a bioprospecting potential.Table No.2 gives a list of disappearing plants plants of the region which are to be conserved , because they are having a bio-prospecting potential. The conservation of biodiversity particularly, the plants which are having bio-prospecting potential and they are discovered from the ethno-medicinal sources are to be protected seriously. The ethnic people providing the knowledge about these plants should be given proper economic share and also trained for the conservation of the medicinal plants by using ex-situ methods and in-situ methods of conservation.

Table 1: List of Some Medicinal plants of Marathwada region of Maharashtra.

S. No.	*Name of plant*	*Family*	*Comman Name*
1.	*Abrus precatorius* L.	Fabaceae	Gunj
2.	*Abutilon indicum* (L.) sweet	Malvaceae	Shikka
3.	*Acalypha indica* L.	Euphorbiaceae	Khokli
4.	*Annona squamosa* L.	Annonaceae	Sitaphal
5.	*Bacopa monnieri* (L.) Wettst.	Scrophulariaceae	Neer brahmi
6.	*Boerhavia diffusa* L.	Nyctaginaceae	Punarnawa
7.	*Boswellia serrata* Roxb.	Burseraceae	Salal

contd...

S. No.	*Name of plant*	*Family*	*Comman Name*
8.	*Calotropis giganeta* (L.)R.Br.	Euphorbiaceae	Runchki
9.	*Careya arborea* Roxb.	Lecythidaceae	Kumbhi
10.	*Cassia tora* L.	Caesalpinaceae	Tarota
11.	*Cissus quadrangular* L.	Vitaceae	Kand wel
12.	*Cocculus hirsutus* (L.) Diels	Menispermaceae	Wasan wel
13.	*Curculigo orchioides* Gaerth	Hypoxidaceae	Kali musli
14.	*Crateva adansonii* Dc.	Capperdiaceae	Waiwarna
15.	*Cuscuta reflexa* Roxb	Cuscutaceae	Amarwel
16.	*Datura inoxia* Mill.	Solanaceae	Dhotra
17.	*Datura metel* L.	Solanaceae	Kala-Dhotra
18.	*Desmodium gangeticum* (L.)Dc.	Fabaceae	Salwan
19.	*Dioscorea bulbifera* L.	Dioscoreaceae	Dukkar kand
20	*Eclipta alba* (L.) Hasssk.	Asteraceae	Maka
21	*Euphorbia antiquorum*	Euphorbiaceae	Tridhari nivdung
22.	*Evolvulus alsinoides* L.	Convolvulaceae	Vishnukanta
23	*Gloriosa superba* L.	Liliaceae	Kal lawi
24.	*Glossocardia boswallea* (L.f.) DC.	Astoraceae	Khadak shepu
25.	*Hybanthus enneaspermus* (L.) Muell.	Violaceae	Ratan purush
26.	*Holarrhena antidysentrica* Sensu. Wall.	Apocynaceae	Pandhra Kuda
27.	*Ipomoea nil* (L.)Roth.	Convolvulaceae	Kala-dana
28.	*Ixora pavetta* Andr.	Rubiaceae	Lokhandi
29.	*Lagerstroemia indica* L.	Lythraceae	Moi
30.	*Leonotis nepetifolia* (L.)R.Br.	Lamiaceae	Deep mal
31.	*Lepidagathis hamiltoniana* Wall.	Acanthaceae	Gundra
32.	*Martynia annua* L.	Martyniaceae	Wagh nakhi
33.	*Mucuna pruriens* DC.	Fabaceae	Khas koiri
34.	*Merremia dissecta* (Jacq.)Hall.f.	Convolvulaceae	Afumari
35.	*Ocimum basilicum* L.	Lamiaceae	Tulas
36.	*Operculina turpethum* (L.)Silva	Convolvulaceae	Nishottar
38.	*Oxalis corniculata* L.	Oxalidaceae	Ambushi
39.	*Phyla nodifloria* (L.) Green	Verbenaceae	Jal Pimpali

contd...

S. No.	*Name of plant*	*Family*	*Comman Name*
40.	*Phyllanthus emblica* Gaerth.	Euphorbiaceae	Awla
41.	*Psoralia corylifolia* L.	Fabaceae	Bawchi
42.	*Semecarpus anacardium* L.f.	Anacardiaceae	Bibba
43.	*Solanum virginianum* L.	Solanaceae	Bhui ringani
44.	*Soymida febrifuga* (Roxb.)A. Juss.	Meliaceae	Rohan
45.	*Sterculia urens* Roxb.	Sterculiaceae	Kondol
46.	*Tinospora cardifolia* (Willd.)Miers.	Menispermaceae	Gulwel
47.	*Woodfordia fruticosa* (L.)Kurz.	Lythraceae	Dhaiti
48.	*Wrightia tinctoria* (Roxb.) R. Br.	Apocynaceae	Kala Kuda
49.	*Xanthium strumarium* L.	Asteraceae	Landga
50.	*Ziziphus rotundifolia* Lam.	Rhamnaceae	Bor

Table 2 : List of disappearing Medicinal plants from Marathwada region which are to be Conserved.

Sr. No.	*Name of Plant*	*Family*	*Medical use*
1.	*Boswellia serrata* Roxb.	Burseraceae	Arthritis
2.	*Crateva adansonii* Dc Subsp.	Capparidiaceae	Kidney stones and Stomach problem
3.	*Curculigo orchioides* Garten	Hypoxidaceae	Liver problems.
4.	*Holarrhena antidysentrica* Sensu	Apocynaceae	Stomach Disorders.
5.	*Merremia dissecta* (L.) Urb.	Convolvulaceae	Antidiabetic
6.	*Mucuna pruriens* Dc.	Fabaceae	Aphrodisic
7.	*Psoraelea corylifolia* L.	Fabaceae	Skin-diseases
8.	*Soyamida febrifuga* (Roxb) A. Juss.	Meliace	Wound healing properties
9.	*Sterculia urens* Roxb.	Sterculaceae	Diabetes and Gynic problem.
10.	*Wrightia tinctoria* Br.	Apocynaceae	Jaundice

REFERENCES

1. May R.M. (2002) " The future of Biological diversity in Crowded World". *Current Sciences* Vol.82 No. 11
2. Rajasekharan P.E.,Ganeshan S. AND Sunitha Bhaskaran (2001) "Conservation of Endangered Medicinal Plants: Challenges and options" *Indian J. Plant Genet. Resour* 14: 296-297
3. Naik V.N. (1998) "*Flora of Marathwada*" Vol No. I & II Amrut Prakashan Aurangabad .
4. Strenlof K. (2000) "*Bioprospecting*" Cold fuel economic incentives for biological conservation, Columbia University news.

Floral Diversity and their Conservation (2013), Editors: D.R. Khanna et al.
Published by Biotech Books.
ISBN: 978-81-7622-286-0 Pages: 239-244

25

Evaluation of Botanicals in Storage against Khapra Beetle (*Trogoderma granarium* everts) and Grain Qualities of Wheat (*Triticum aestivium* cv. HD2329)

SANJIV CHARJAN[1], PARIKSHIT BOKARE[2], AMOL PATIL[2], RAVIRAJ UDASI[2], ASHISH LAMBAT[3], RAJESH GADEWAR[3]
[1]Assistant Professor, College of Agriculture (Dr. PKV's), Nagpur
[2]Student, College of Agriculture (Dr. PKV's), Nagpur
[3]Sevadal Mahila Mahavidyalaya and Research Academy, Nagpur, Maharashtra, India.

The khapra beetle (*Trogoderma granarium* Everts) is a serious store grain pest of wheat during storage. In the present investigation, the experiment was conducted to know the effect of different organic grain protectants on infestation percentage of khapra beetle and seed qualities of wheat (*Triticum astivium* cv. HD2329). It was observed that seed treated with sweet flag powder (2.5%) and custard apple seed powder 2.5% were showed significantly higher 100 seed weight, germination percentage, seedling vigour, field emergence percentage and adult mortality as compared to other seed treatment and controll during storage.

Key words: organic grain protectants, sweet flag, custard apple, khapra beetle, wheat infestation, storage.

INTRODUCTION

The khapra beetle (*Trogoderma granarium* Everts) is a serious pest of stored grain causing considerable damage to almost all cereals in storage. It is largely responsible for damage and frequently harboring stores, mill and warehouses. Khapra beetle is cosmopolitan in nature attributing about 50% loss in seed weight during storage.

The steady rise in the use of pesticides for control of store grain pests can be dangerous to human being and cattle as well due to their residual toxicity. With view to find out safe and organic seed protectants, present investigation was taken up to evaluate the organic grain protectants in seed storage against khapra beetle in wheat.

MATERIALS AND METHODS

Wheat (*Triticum aestivium* cv. HD2329) seed were used in various phages of this study, produced in 2008-2009. The seed were cleaned and dried (moisture content 10.5%). The wheat seed were treated (may 2009) with six plant product *viz.*, Neem leaf, sweet flag, Tulsi, Custard apple, Turmeric and Pongamia powder is in the proportion of 2.5% by weight of the seeds. The experiment as conducted in glass bottle of 1 lit capacity with seven treatments including untreated control. Each glass bottle was then filled with 500 gm of wheat seeds. 10 pairs of 2-3 days old khapra beetle were released in each glass bottle covered with muslin cloth. The set of experiment was kept in well ventilated wire mesh almirah in masonry building having cemented walls, roof and floor under ambient temperature (24.1-45.8°C) and relative humidity (20-85%)from may to july,2009. After three months the seed from each treatment were keenly observed and those found infested were separated out weighted to determine the infestation percentage on weight basis hundred seed weight and germination were tested in quadruplicate with 100 seed in each replication. The germination percentage was evaluated on the value for normal seedling (Anon 1985). the vigour index were workout following the method of Abdul Baki and Anderson (1973). For field emergence test, sowing of wheat seed was done in randomized block design, with four replication with inter and intra row spacing of one feet and six inches respectively. Observations for field emergence were recorded daily and finally the established seedlings were counted after one month of sowing. The experimental data was statistically scrutinized as per the Panse and Sukhatme (1967).

RESULT AND DISCUSSION

The data regarding the effect of the different organic grain protectants on population behavior (Adult mortality), infestation percentage 100 seed weigh, germination percentage, vigour index and field emergence percentage after three months of storage are given in table 1.

Table 1: The Effect of different organic protectants on khapra beetle mortality, infestation percentage, 100 seed weight, germination percentage, seedling vigour and field emergence percentage.

Treatments	*Adult Mortality (%)*	*Percent weight loss due to infestation (%)*	*100 seed weight*	*Germination (%)*	*Seedling vigour index (SVI)*	*Field emergence (%)*
Neem leaf (2.5%)	60	5.8	3.02	69	1338	57
Sweet flag (2.5%)	100	0.05	3.32	96	2018	88
Tulsi (2.5%)	51	6.4	2.91	65	1291	55
Custard apple (2.5%)	92	2.4	3.17	93	2006	84
Turmeric (2.5%)	47	6.8	2.85	61	1124	50
Pongamia (2.5%)	56	6.0	2.97	64	1202	53
Untreated control	5	17.14	2.01	35	682	22
SEm(±)	0.35	0.09	0.03	0.12		0.10
CD at 5%	1.06	0.29	0.10	0.36		0.31

The result indicated that variation in number of khapra beetle (adult) in each treatment. The khapra beetle adult mortality was significantly highest in wheat seed treated with sweet flag(100%) which is closely followed by custard apple (92%) neem leaf (60%), pongamia (56%), tulsi (51%) and turmeric (47%). Where in significantly lower mortality was observed in untreated control (5%). Saxena et. al (1976), Tikku *et al.* (1978) and Khan and Borle (1985) found *Acorus calamus* L oil vapour responsible for causing infecundity among the female of a number stored grain pest. Biradar (2000) who reported that sweet flag has got insecticidal and ovicidal effect. The significantly wheat seed weight loss was observed in untreated control followed by turmeric, tulsi, pongamia, neem leaf, custard apple and sweet flag treatment during entire period of storage. This might be due to sterilizing effect of sweet flag rhizome powder mixed with wheat seeds. Charjan and Tarar (1994), Deshpande *et al.* (2010), Ashish *et al.* (2011), and K. Cherian *et al.* (2011) reprted that the sharp declined in infestation percentage of store grain pest in seed treated with *Acorus calamus* powder.

The seed quality parameters *viz.*, 100 seed weight, germination percentage and seedling vigour index was highest in seed treated with 2.5% concentration of sweet flag powder followed by custard apple, neem leaf, pongamia, tulsi, turmeric and untreated control. The 100 seed weight, germination percentage and vigour index decreased with increasing infestation of stored grain pest (Howe, (1972), Charjan and Tarar (1994), Deshpande *et al.* (2010) and Ashish Lambat et. al (2011). Since the stored grain pest have been eaten off major portion of the endosperm which leads to reduction in weight of the wheat seeds and in turn affect the seed germination and vigour index because of lack of stored food and is in conformity with the findings of Narayanaswami (1985). Handerson and Christensen (1961) reported that pulse beetle attack the embryo and germination potential of seed reduced or totally destroyed.

The field emergence percentage of wheat seeds follow the same trends of seed quality parameters. The field emergence percentage was highest in seed treated with 2.5% concentration of sweet flag powder as compared to other treatments and untreated control. This might be due to the least infestation of khapra beetle and higher 100 seed weight, germinability and seedling vigour index. The results are agreement to those reported by Charjan and Tarar (1994), Deshpande *et al.* (2010), Ashish Lambat et. al (2011) and K. Cherian *et al.* (2011).

Among the plant products sweet flag powder 2.5% were found to be ignificantly effective against khapra beetle through out the period of investigation these findings are agrriment with Siva Srinivasu (2001), Deshpande *et al.* (2010), Charjan and Tarar (1994) and Ashish Lambat et. al (2011) thus sweet flag and custard apple naturally occurring botanicals which are not toxic can be used as pre storage seed treatment, dispensing with the use of costly and toxic chemicals to control khapra beetle damage without adversely affecting the germination of wheat seed.

REFERENCES

1. Abdul Baki, A.A. and Anderson J.D. 1973 Vigour determination in soybean seed by multiple criteria. Crop sci. 13: 630-633.
2. Anonymous, 1985 International rules for seed testing. Seed sci. and Technol. 13: 299-513
3. Ashish Lambat, Rajesh Gadewar, Sanjeev Charjan, Konglath Cherian and Prachi Lambat, 2011. Evaluation of organic grain protectants in seed storage against rice weevil in wheat. Proc. of International conference on Sustainable Environment held during 19-20 February 2011 at Aurangabad, Maharashtra, India. Special issue 6: 122-123.
4. Biradar, B.S. 2000 Prevention of cross infestation by *Sitophilus oryzae* L. and *Rhizopertha dominica* in stored wheat. M.Sc. (Agri.) Thesis, University of Agricultural Sciences, Dharwd.
5. Charjan S.K.U. and Tarar, J.L. 1994 The influence of some plant products on seed quality of lobia during storage. Ann. Plant Physiol. 8(2)L: 153-156.
6. Deshpande, V.K. Deshpande, H.H. and Masuthi, D. 2010 Evaluation of grain protectants in seed storage against *Sitophilus oryzae* (L.) in sorghum. Green farming 1(5):512-514.
7. Handerson, L.S. and Christensen, C.M. 1961 Preharvest control of insect and fungi. U.S. Dept. Agri. Ybk, pp. 348-356.
8. Howe, R.W. 1972 Insect attacking seed during storage. Seed Biology vol. III (ed. Kozlowski, T.T.) Academic press: New York pp. 247-300.
9. Khan, M.I. and Borle, M.N. 1985 Efficacy of some safer grain protectants against the pulse beetle, Collosobruchus chenensis L. infecting stored Bengal gram (*Cicer arietinum* L.). P.K.V. Res. J. 9(1): 53-55.
10. Konglath Cherian, Sanjeev Charjan, Vandan Mohod, Ashish Lambat and Rajesh Gadewar 2011. Studies on the influence of *Acorus calamus* L. rhizome Powder seed treatments against stored grain pest of wheat. Proc. of National Seminar on Environmental Management and Biodiversity conservation, held during 26-27 February 2011 at Rishikesh (U.K.) India. Abstract no A127 page 101.
11. Narayanaswamy, S 1985 Effect of Pulse beetle damage on seed quality of field bean and Pigeon pea. Seed Res. 13(2): 138-141.
12. Panse, V.G. and Sukhatme, P.V. 1967. Statistical methods for agricultural workers. I.C.A.R. Pub., New Delhi .

13. Saxena, B.P., Koul, O. andTikku, K. 1976. Non toxic protectants against the stored grain insect pest. Bull. Grain Technol. 14(5): 190-193.
14. Tikku K., Koul,O. and Saxena B.P. 1978. The influence of *Acorus calamus* L. oil vapour on the histocytological pattern of the ovaries of *Trogoderma granarium* Evert. Bull. Grain Tehnol. 16 (1): 3-9.

Floral Diversity and their Conservation (2013), Editors: D.R. Khanna et al.
Published by Biotech Books.
ISBN: 978-81-7622-286-0 Pages: 245-258

26

Native Flora of Various Compartments of Gorewada Forest, Nagpur District, Nagpur (M.S.), India

Kirti V. Dubey[1*], Sulbha V. Kulkarni[1], Pravin Charde[1]
[1] Sevadal Mahila Mahavidyalaya, Sakkardara Square,
Umrer Road, Nagpur, India

Gorewada forest is one of the tropical forest located in Nagpur district, (M.S.), India. It is a rich source of biodiversity which represents valuable global resources such as food, fodder, fruits, fuel, gums and medicines that are beneficial to mankind. Among various resources of forests, legumes and non-legumes represents a very large and diverse group of plant kingdom, ranging from herbs, shrubs, trees, climbers and grasses. In the present study, variety of legume and non-legume plants from seven different compartments of Gorewada forest *viz.* 790-796 of Gorewada forest of Nagpur district were surveyed and collected for taxonomic identification of plants and their distribution pattern *i.e.* density to find out the dominant flora. Most of the plants in the different compartments are common which includes the leguminous plants which were in large number as compared to non-legumes. Leguminous plants found are *Abrus precatorius, Acacia* sp., *Albizia lebbeck, Butea monosperma, Bahunia* sp., *Caesalpinia pulcherima, Cassia fistula, Cassia tora, Leucaena leucocephala, Mimosa hamata,*

Mucuna pruriens, Pithecellobium dulce, Pongamia pinnata, Tamarindus indica, Tephrosia purpurea, Tephrosia hamiltonii etc. and were as non-legumes are *Blumea* sp., *Tectona grandis, Santalabum album, Zizipus* sp., *Celosia argentea, Cleistanthus collinus, Mytragyna perbifolia, Partheneium* sp., *Semicarpus anacardium, Sida* sp., *vernonia cinerea* etc. This study will be useful in determining the species diversity to characterize the structure of the community available in and around Gorewada International Zoo. The plant diversity study of the Gorewada forest is the first of its kind and the diversity indices calculated were viz Simpson's index (D=0.343), Simpson's index of diversity (1-D= 0.657), Simpson's reciprocal index (1/D = 2.9154), Shannon-wiener index (H= 1.204) and Evenness index (E= 0.743)

Key words: Gorewada forest, biodiversity, Simpson's index, Shannon-Wiener index, Evenness index.

INTRODUCTION

The forest cover in India was estimated as 675,538 sq. km. and constitutes 20.55 percent of its geographic area. Of this, dense and open forest cover constitutes as 12.68 and 7.87 percent, respectively. However, the area of non-forest in the country is 2,611,725 sq. km. and constitutes 79.45 percent of the geographical areas (1). Development of richness in the vegetation wealth and biodiversity of soil is of prime importance in the forest cover as well as on the non - forest area to conserve the bioresource and tree wealth.

Forests of the tropical zones constitutes about half of the world's forest and mostly occur in developing countries. In recent years tropical forests have received much attention because of their species richness, high standing biomass, and global net primary productivity. Tropical forests consist of world's largest biodiversity and play an important role in the global terrestrial carbon budget. The structure, composition and functioning of this forests are undergoing rapid changes because of anthropogenic activities, biotic pressure and widespread economic growth are altering the natural vegetal cover and putting tremendous pressure on the sustenance of the few left over tropical forest covers in India. As a result, there is lot of spatial and temporal variations in the reported values of species richness, composition and productivity. There is a pressing need to monitor the rate and extent of changes in the tropical forest covers in countries like India for efficient planning and management leading to sustainable development. The loss of natural habitat and species extinctions around the world have served to focus intense international attention on the issue of biodiversity and it is widely recognized that the path to sustainable development requires the conservation and wise use of the world's biodiversity. Indeed, measuring and monitoring biodiversity must be the first step towards effective conservation and sustainable development.

This paper highlights the studies on survey of native plant species in and around each compartment of Gorewada forest which is proposed site for the development of International Zoo, in order to find out the native dominant species of each compartment. Species diversity of the plants available in Gorewada forest was estimated through mathematical measures to characterize the structure of a community and their relative abundance.

Aims & Objectives: Aims of the present study is to characterize the structure of a community of the dominant plant species and their relative abundance in and around each compartment of Gorewada forest, Nagpur (M.S.), India.

MATERIAL AND METHODS

Site survey

To achieve the proposed objectives intensive and extensive site surveys were conducted at Gorewada forest during December-2007 to March-2008 by the college team of microbiologist, botanist and soil scientist to cover the total area of Gorewada forest area *i.e.* 1881.66 ha.

Survey of native plant species in and around Gorewada forest

The tour in and around different compartments of Gorewada forest was made in phased manner so to cover most of the parts of the Gorewada forest for the survey of native plant species. Specimens were brought to the college laboratory and all plants were identified by using reference floras (2-6). For collection and preservation of plant samples, standard procedures were generally followed (7-8). The trips were arranged in such a way so as cover all the compartments of Gorewada forest and collect most of the plants in flowering or fruiting stages. All the specimens collected were serially numbered. The field notes were taken regularly, included habitat, colour of the flowers, association and other pertinent features. Efforts were made to identify the plants from the fresh material; those that could not be satisfactorily identified in the field were brought to the laboratory and identified by checking it with monographs, herbarium specimens and other available literature.

RESULTS AND DISCUSSION

Description of Gorewada forest Site

Intensive and extensive site surveys were conducted at Gorewada forest during December-2007 to March-2008. During survey it was found that seven different compartments has characteristic features with respect to plants species available during winter season. Total area of Gorewada forest area is 1881.66 ha. and the forest area is unequally divided in to two parts by Nagpur-Kalmeshwar highway. Among the two, part on left side of the Nagpur-Kalmeshwar highway, is divided into three compartments allotted with numbers 790, 791, and 792 and second part, to the right side of the Nagpur-Kalmeshwar highway, has four compartments allotted with No.793, 794, 795 and 796 along with a water tank of 57.93 ha.

Description of Flora available in different Compartments of Gorewada Forest

Compartment No. 790:

This site has almost a barren look with very few plant species towards its entrance with some of the pits dug for plantation on barren land. However, the density of plants improved towards the center. Types of plant species comprising of trees, shrubs, herbs, climbers and grasses present in this compartment is presented in Table 1. Among the plant species available, *Alternanthera pungens* and *Cassia tora* (herbs) and trees such as *Acacia catechu, Butea monosperma, Cassia siamea, Diospyros melanoxylon, Maytenus emarginata, Mitragyna perbifolia, Ziziphus mauritiana* are the predominant species followed by *Acacia catechu, Albizia odoratissima, Azadirachta indica* and *Daulbergia sisso*. Along the extremity of the compartment *i.e.* towards Nagpur-Kalmeshwar highway monoseries plantation of *Dendrocalamus strictus* was found which was planted to provide protective barriers from invaders. The forest present in 790 compartment is of dry deciduous type.

Compartment No. 791:

This site also has almost a barren and dry deciduous look like that found in compartment No. 790. with comparatively a very few plant species. However, the density of plants improved towards the center. Types of plant species comprising of trees, shrubs, herbs, climbers and grasses present in this compartment is presented in Table 1. Among the plant species available, *Alternanthera pungens, Cassia tora* (herbs) and *Acacia catechu, Butea monosperma, Mitragyna perbifolia, Cassia siamea, Maytenus emarginata, Ziziphus mauritiana, Diospyros melanoxylon* (trees) are the predominant species followed by *Daulbergia sisso, Albizia odoratissima, Acacia catechu* and *Azadirachta indica.*

Compartment No. 792:

In compartment number 792 of Gorewada forest about 42 different plant species classified as trees, shrubs, herbs, climbers and grasses (**Table 1**). Among the tree species dominant types were *Acacia catechu, A. leucophloea. A. nilotica, Albizia odoratissima, Butea monosperma, Cassia siamea, Mitragyna perbifolia, and Ziziphus mauritiana.*

Compartment No. 793:

This compartment is located at West side of Gorewada Catchments area. Multi-species plantation was found which comprised of predominant species *viz. Acacia catechu, Hardwickia binata*, *Butea monosperma*, *Cleistanthus collinus*, *Dalbergia sisso* followed by *Santallum album*, *Acacia arabica*, *Bombox cieba*, *Acacia leucophloea*, *Soymida febrifuga*, *Mimosa hamata*, *Ziziphus mauritiana* (Table 1). Different types of herbs and shrubs were also found. Fruit bearing plants of *Ziziphus oenoplia* and *Abrus precatorius* were also found. Plantation of S*antalum album* was also observed

Compartment No.794:

This compartment is situated on the right side of Nagpur - Kalmeshwar highway. Major part of this compartment is densely populated with multiple plant species (Table 1). However, successful monospecies plantation of *Xylia xylocarpa* is found in small area of this compartment with few plants of *Butea monosperma*, and *Azadirecta indica.* Perennial plant species of this compartment are *Maytenus emarginata*, *Butea monosperma, Acacia catechu, Acacia nelotica, Delbergia sisso, Cassia fistula, Ailanthus excels, Acacia leucophloea, Albizia lebbeck, Pithecellobium dulci, Pongamia pinnata, Gmelina arborea, Phoenix sylvestris, Dendrocalamus strictus* along with herbs, shrubs and climbers. Different types of thorny shrubs *Ziziphus rotandifolia* bearing fruits are found in this dry deciduous compartment. Termite infestation was found in *Butea monosperma* and *Acacia leucophloea.* An example of climber *viz. Macuna pruriens* present in compartment number 794. Barren patch of land is also found in compartment number 794.

Compartment No.795:

This compartment has a water body measuring 57.93 ha. and multispecies plantation and list of plants found during survey is presented in Table 1. Some of the predominant types of plant species available in compartment number 795 are *Acacia catechu, Acacia leucophloea, Butea monosperma, Cleistanthus collinus, Mimosa hamata, Tectona grandis,* and *Ziziphus mauritiana.*

Compartment No. 96:

This compartment is adjacent to Compartment number 794. It is characterized to have natural hilly area. In the centre of this compartment almost barren rocky area with very few plant species *viz., Tectona grandis, Ziziphus mauritiana* and *Ziziphus oenoplia* are found giving dry deciduous look to the compartment. The barren area of this compartment needs plantation. However, on the lower side of hilly area there is Gorewada reservoir and on top of the hilly area dense forest is present. Top of the hilly area is densely covered with green canopy and the dominant plant species available are *Maytenus emarginata, Ziziphus oenoplia, Butea monosperma, Acacia catechu, Acacia nelotica, Delbergia sisso, Cassia fistula, Ailanthus excelsa, Acacia leucophloea, Albizia lebbeck, Pithecellobium dulci, Pongamia pinnata, Gmelina arborea, Phoenix sylvestris, Dendrocalamus strictus, Tectona grandis, Semicarpus anacardium, Mimosa hamata, Grewia tiliifolia, Hardwickia binata, Terminalia arjuna* (Table 1). In this compartment there is mother plant of *Terminalia arjuna.* In this compartment, barren patch of land is present which is almost devoid of vegetation due to inhabitation of Sita Gondi people (the local tribal community). However, *Ziziphus* sp. were found sparsely. Amanalla is passing through this compartment that is a centre of Gorewada catchment area. Few plants of *Tectona grandis* and *Acacia leucophloea* are found to be infested by termites.

Diversity Indices of Plants available at Gorewada forest

It is a measured by a number of diversity indices. There has been a lot of development and proliferation of diversity indices over the years. These indices are mathematical expressions that combine three components of community structure, *i.e.* richness (number of species present), evenness (the distribution of individuals among species), and abundance (total number of plants present). The plant diversity studies of the Gorewada forest is the first of its kind carried out and results are presented in Table 2. The diversity indices, Simpson's index (D=0.343): Simpson's index is a dominance measure and a probability measure to study forest plant biodiversity (9, 10). The higher value at Gorewada forest could mean the high dominance of the occurring species. There were trees higher in number followed by herbs, shrubs, climbers and grasses. Index of similarity (0.657) indicates high diversity. Reciprocal index was 2.9154. Shannon-wiener index was 1.204 indicating quite low diversity, which is lower than the reported range of 1.5 to 3.5 (11). Finally, Evenness index was 0.743 which indicates that abundance of different species is similar in proportion to all species.

CONCLUSION

Present study was carried out to find native as well as dominant plant species in and around each compartment of Gorewada forest, a proposed site for the development of International Zoo. Species diversity of the plants available in Gorewada forest was estimated through mathematical measures to characterize the structure of a community and their relative abundance. Variety of legume and non-legume plants from seven different compartments of Gorewada forest *viz.* 790-796 of Gorewada forest of Nagpur district were surveyed and collected for taxonomic identification of plants and their distribution pattern *i.e.* density to find out the dominant flora. Most of the plants in the different compartments are common which includes the leguminous plants which were in large number as compared to non-legumes. The plant diversity study of the Gorewada forest is the first of its kind and the diversity indices calculated were viz Simpson's index (D=0.343), Simpson's index of diversity (1-D– 0.657), Simpson's reciprocal index (1/D = 2.9154), Shannon-wiener index (H= 1.204) and

Table 1: List of Various Plant Species Present in different Compartments of Gorewada Forest

S. No	**Various compartments with different plant species**						
	No.:-790	**No.:-791**	**No.:-792**	**No.:-793**	**No.:-794**	**No.:- 795**	**No.:- 796**
	Trees	**Trees**	**Trees**	**Trees**	**Trees**	**Trees**	**Trees**
1	*Acacia catechu*	*Acacia catechu*	*Acacia catechu*	*Acacia arabica*	*Acacia catechu*	*Acacia catechu*	*Acacia catechu*
2	*Acacia leucophloea*	*Acacia leucophloea*	*Acacia leucophloea*	*Acacia catechu*	*Acacia leucophloea*	*Acacia leucophloea*	*Acacia leucophloea*
3	*Acacia nilotica*	Acacia nilotica	*Acacia nilotica*	*Acacia leucophloea*	*Acacia nilotica*	*Azadirachta indica*	*Acacia nilotica*
4	*Adina cordifolia*	Aegel marmalosa	Albizia odoratissima	*Acacia nilotica*	Ailanthus excels	*Butea monosperma*	*Azadirachta indica*
5	*Ailanthus excelsa*	Ailanthus excelsa	Andropogon pumilus	*Albizia lebbeck*	Albizia lebbeck	*Cleistanthus collinus*	*Bauhinia purpuria*
6	*Albizia odoratissima*	*Albizia odoratissima*	Azadirachta indica	Albizia odoratissima	Azadirachta indica	*Emblica officinalis*	*Bauhinia racemosa*
7	*Anogeissus latifolia*	*Azadirachta indica*	Bauhinia racemosa	Annona squamosa	*Bauhinia racemosa*	*Leucaenea leucocephala*	Bombaxi cieba
8	*Azadirachta indica*	*Bauhinia racemosa*	*Butea monosperma*	*Azadirachta indica*	*Butea monosperma*	*Mimosa hamata*	*Butea monosperma*
9	*Bauhinia racemosa*	*Butea monosperma*	*Cassia siamea*	*Bombaxi cieba*	*Cassia fistula*	*Phoenix sylvestris*	*Dalbergia sissoo*
10	*Bombaxi cieba*	*Cassia siamea*	*Dalbergia sissoo*	*Butea monosperma*	*Dalbargia sissoo*	Pongamia pinnata	*Dandrocalamus strictus*
11	*Butea monosperma*	*Dalbergia sissoo*	*Dandrocalamus strictus*	*Cleistanthus collinus*	*Dandrocalamus strictus*	*Santalum album*	*Delonix regia*

contd...

12	*Cassia angustifolia*	Dandrocalamus strictus	*Dolichandrone falcata*	*Dalbergia sissoo*	*Leucaena leucocephala*	*Tamarindus indica*	*Diospyros melanoxylon*
13	*Cassia siamea*	Delonix regia	*Maytenus emarginata*	*Eucalyptus sp.*	*Maytenus emarginata*	*Tectona grandis*	*Grewia tiliifolia*
14	*Dalbargia sissoo*	*Diospyros melanoxylon*	*Mimosa hamata*	*Gemelina arborea*	*Mimosa hamata*	*Ziziphus mauritiana*	*Leucaena leucocephala*
15	*Dandrocalamus strictus*	*Leucaena leucocephala*	*Mitragyna perbifolia*	*Hardwickia binata*	*Mitragyna perbifolia*	Shrubs	Maytenus emarginata
16	*Diospyros melanoxylon*	*Maytenus emarginata*	*Phoenix sylvestris*	Leucaenea leucocephala	Phoenix sylvestris	Lantana camera	Mimosa hamata
17	*Dolichandrone falcata*	*Mimosa hamata*	*Semicarpus anacardium*	Merremia emarginata	*Pithecellobium dulce*	*Xanthium strumarium*	*Mitragyna perbifolia*
18	*Emblica officinalis*	Mitragyna perbifolia	Tamarindus indica	*Mimosa hamata*	*Pongamea pinnata*	Climbers	Phoenix sylvestris
19	*Grewia tiliifolia*	*Phoenix sylvestris*	*Tectona grandis*	*Phoenix sylvestris*	*Tamarindus indica*	*Abrus precatorius*	*Pithecellobium dulce*
20	*Leucaena leucocephala*	*Pongamia pinnata*	Terminalia bellirica	*Santalum album*	*Tectona grandis*	*Ipomea aquatic*	*Santalum album*
21	*Maytenus emarginata*	*Semicarpus anacardium*	Ziziphus glaberrima	Soymida febrifuga	*Woodfordia fruticosa*	*Ipomea quamoclit*	Semicarpus anacardium
22	*Mimosa hamata*	*Tamarindus indica*	Shrubs	Ziziphus glaberrima	*Xylia xylocarpa*	Herbs	Sesamum orientale
23	*Mitragyna perbifolia*	*Terminalia bellirica*	*Lantana camera*	Ziziphus mauritiana	*Ziziphus mauritiana*	*Ageratum conyzoides*	*Tamarindus indica*
24	*Phoenix sylvestris*	Ziziphus glaberrima	Xanthium strumarium	Shrubs	Shrubs	*Alternanthera pungens*	Tectona grandis
25	*Pithecellobium dulce*	Ziziphus mauritiana	Climbers	*Goniocalon glabrum*	*Ipomea aquatic*	*Blumea eriantha*	*Terminalia alata*

contd...

26	Semicarpus anacardium	Shrubs	Cocculus hirsutus	*Lantana camera*	*Lantana camera*	Cassia tora	Woodfordia fruticosa
27	*Soymida febrifuga*	*Lantana camera*	*Coix lacryma-jobi*	*Xanthium strumarium*	*Xanthium strumarium*	*Goniocalon glabrum*	*Ziziphus glaberrima*
28	*Tamarindus indica*	*Xanthium strumarium*	Hemidesmus indicus	*Ziziphus oenoplia*	*Ziziphus oenoplia*	*Hyptis suaveolens*	*Ziziphus mauritiana*
29	*Tectona grandis*	Climbers	*Ipomea quamoclit*	*Ziziphus xylopyrus*	Climbers	*Ocimum sanctum*	Shrubs
30	*Terminali bellirica*	*Cocculus hirsutus*	Herbs	Climbers	*Combretum ovalifolium*	*Parthenium*	*Cappris zeylanica*
31	*Terminalia alata*	*Hemidesmus indicus*	*Ageratum conyzoides*	*Hemidesmus indicus*	*Cuscuta reflexa*	*Sida acuta*	*Lantana camera*
32	*Ziziphus glaberrima*	*Ipomea quamoclit*	*Alternanthera pungens*	*Ipomea quamoclit*	*Hemidesmus indicus*	*Sida cordata*	*Malachra capitata*
33	*Ziziphus glaberrima*		*Blumea eriantha*	*Mucuna pruriens*	*Mucuna pruriens*	*Tridax procumbens*	Climbers
34	*Ziziphus mauritiana*		Cassia tora	Herbs	Herbs	*Vernonia cinerea*	*Cocculus hirsutus*
	Shrubs		*Hyptis suaveolens*	*Ageratum conyzoides*	*Alternanthera pungens*	*Vicoa indica*	*Coix lacryma-jobi*
35	*Lantana camera*		Parthenium	*Alternanthera pungens*	*Blumea eriantka*	*Vinca rosea*	*Combretum ovalifolium*
36	*Xanthium strumarium*		*Sida acuta*	*Blumea erintha*	*Cassia tora*	Grasses	*Hemidesmus indicus*
	Climbers		*Sida cordata*	*Cassia tora*	*Celosia argentea*	Cynodon dactylon	*Ipomoea aquatic*
37	*Cocculus hirsutus*		*Tridax procumbens*	*Celosia argentea or C. Cristata*	*Cyathocline purpuria*	*Hackelochloa granularis*	*Mucuna pruriens*

contd...

38	*Combretum ovalifolium*		*Vernonia cinerea*	*Hyptis suaveolens*	*Hyptis suaveolens*		Herbs
39	*Hemidesmus indicus*		*Vicoa indica*	*Parthenium hysterophorus*	*Malachra capitata*		*Alternanthera pungens*
40	*Mucuna pruriens*		Grasses	*Sida acuta*	*Parthenium hysterophorus*		*Blumea eriantha*
	Herbs		*Apluda mutica*	*Sida cordata*	*Sida acuta*		*Cassia tora*
41	*Alternanthera pungens*		*Hackelochloa granularis*	*Tridax procumbens*	*Sida cordata*		*Celosia argentea*
42	*Blumea eriantha*		*Iseilema laxum*	*Vernonia cinerea*	*Solanum xanthocarpum*		*Cyathocline purpuria*
43	*Cassia tora*			*Vicoa indica*	*Tephrosia hamiltoni*		*Hyptis suaveolens*
44	*Celosia argentea*				*Tephrosia purpureai*		*Parthenium hysterophorus*
45	*Cyathocline purpuria*				*Tridex procumbens*		*Pennisetum hohenackeri*
46	*Hyptis suaveolens*				*Vernonia cinerea*		*Sida acuta*
47	*Malachra capitata*				Grasses		*Sida cordata*
48	*Parthenium hysterophorus*				*Andropogon pumilus*		*Solanum xanthocarpum*
49	*Sida acuta*				*Apluda mutica*		*Tephrosia purpureai*
50	*Sida cordata*				*Coix lacryma-jobi*		*Tridex procumbens*
51	*Solanum xanthocarpum*				*Eragrostiella bifaria*		*Typha angustata*

contd...

52	*Tephrosia hamiltoni*				*Iseilema laxum*		*Vernonia cinerea*
53	*Tephrosia purpureai*				*Pennisetum hohenackeri*		*Xanthium strumarium*
54	*Tridex procumbens*						*Grasses*
55	*Vernonia cinerea*						*Andropogon pumilus*
	Grasses						*Apluda mutica*
56	*Andropogon pumilus*						*Iseilema laxum*
57	*Apluda mutica*						
58	*Coix lacryma-jobi*						
59	*Eragrostiella bifaria*						
60	*Iseilema laxum*						
61	*Pennisetum hohenackeri*						

Evenness index (E= 0.743). This study will be useful in determining the species diversity to characterize the structure of the community available in and around Gorewada International Zoo.

ACKNOWLEDGEMENTS

Authors acknowledge Director, Gorewada forest, Department of Forest, Government of Maharashtra, Nagpur, India for providing the financial support to carry out the work presented in this manuscript.

Table 2: Plant Diversity Indices for Gorewada Forest.

Measure	*Value*
Simpson's index (D)	0.343
Simpson's index of diversity (1 – D)	0.657
Simpson's reciprocal index (1 / D)	0.291
Shannon-wiener index (H)	1.204
Evenness index (E)	0.743

REFERENCES

1. Katiyar, S. R. Sustainable management of forest resources and tree wealth in India. In: Biodiversity and conservation. Prof. Arvind Kumar edited, p.p. 81-91, (2005).
2. Hooker, K. K. The Flora of British India. Vol. I – VII London, (1872 - 1897).
3. Ugemuge, N. R. Flora of Nagpur district. Shree Prakashan, Nagpur, (1986).
4. Almeida, S.M. The flora of Savantwadi. Vol. I–II. Scientific Publishers, Jodhpur - India., (1990).
5. Joshi, S. J Medicinal plants. Oxford and IBH publishing company Pvt. Ltd., (2000).
6. Singh N. P., Narsighan L. kartikeyan S. and Prasanna P.V. Flora of Maharashtra. BSI Publications, Kolkata, (2001).
7. Jain & Rao A Handbook of field and Herbarium methods. Today & Tomorrow's printers & Publishers, New Delh, (1977).

8. Balgooy. M. M. J. Van. A plant geographical analysis of Sulawesi. In: TC Whitmore (ed.). Biogeographical evolution of the Malay Archipelago. pp. 94-10, (1987).

9. Magurran E. Anne. Ecological diversity and its Measurements. Princton University Press, 41 Williams street, Princeton, New Jersey, (1988)

10. Sai V. S. and Mishra M. Comparison of some indices of species diversity in the estimation of the actual diversity in a tropical forest: A case study. Trop. Ecol. 27: 195-201, (1986)

11. Margalef R. Homage to Evelyn Hutchinson or why is there an upper limit to diversity. Trans. Connect. Acad. Arst. Sci. 44: 211-235, (1972).

Floral Diversity and their Conservation (2013), Editors: D.R. Khanna et al.
Published by Biotech Books.
ISBN: 978-81-7622-286-0 Pages: 259-266

27

Fungal Biodiversity of College Library and Cellulolytic Activity of some Fungi

Dalal L.P.[1], Bhowal M., Kalbende S.P.
Department of Botany, Hislop College, Nagpur
[1]Department of Botany, J. B. College of Science, Wardha

Survey of the college library has brought out valuable information on the allergenic forms. The fungal spore incidence inside a library was recorded by exposing PDA and Czapek dox agar culture medium petriplates for 10 minutes and then incubated at 28 ± 1°C. for 4-5 days and were regularly examined. Numbers of fungal colonies were recorded from two sections of the college library of Wardha city for six months. Petriplate exposure method shows the prevalence of *Aspergillus niger, Aspergillus fumigatus, Aspergillus flavus, Curvularia spp., Alternaria alternata, Rhizopus spp., Chaetomium spp., Penicillium spp., Fusarium spp.* etc. Among the various species encountered *Aspergillus niger, Aspergillus fumigatus* were the dominant fungi in the college library. Investigations by the petriplate exposure method helped us in determining the occurrence of fungal species in the air inside the college library. The cellulolytic activity and damage of library materials by some common airborne fungi were studied.

Key words: PDA, Czapek dox agar, Petriplate exposure method, College library, Wardha city, Cellulolytic activity.

INTRODUCTION

Air is a natural medium for certain very minute particles including many micoflora. Fungal spores constitute a significant fraction of airborne bioparticles [1,2] and they are often 100-1000 times more numerous than other airborne particles like pollen grains [3, 4]. Investigation on aeromycoflora in libraries was carried out in past by many workers [5-11]. Recently only few records were seen highlighting indoor aeromycoflora [12-15]. Airborne mycoflora are largely determined by topography, meteorological parameters, vegetation and biotic factors including human activities [3, 16]. The mycoflora concentrations in the atmosphere are influenced by the processes involved in their production, release and deposition [16]. The present study has been carried out to screen the mycoflora of air inside the library of Wardha city. The study of indoor aeromycoflora of library and fungi associated with biodeterioration of books is important not only for conservation of books but also to prevent diseases that might spread to human beings frequenting the libraries. Some species of *Aspergillus, Penicillium* can cause extreme allergic reaction or respiratory and other related diseases in humans. Libraries have volumes of such suitable substrates in the form of old papers, binding fabrics, glue, dust and air coolers. However, most of the research done pertained to the deterioration of books. The physical condition from different college library structures, such as humidity level, temperature and the presence of organic and inorganic substrates, influence the fungal concentration in their indoor air. Collection of airborne spores can provide valuable information about the indoor air quality in college library.

MATERIAL AND METHODS

Air sampling was carried out in two sections of the college library of Wardha city. One is the reading room and the other where books are stored in racks. Sterilized petri plates of 10 cm diameter containing potato dextrose agar and czapek dox agar were exposed for 10 minutes. Streptomycin was added to inhibit the bacterial growth. The petriplates were incubated at 28±1°C. for 4-5 days. The isolated fungi were examined and identified with the help of authentic literature. This method shows the prevalence of *A. niger, A. Fumigatus, A. flavus, Curvularia* spp., *Alternaria alternata, Rhizopus* spp., *Chaetomium* spp., *Penicillium* spp., *Fusarium* spp. etc. Among the various species encountered *A. niger* and *A. fumigatus* were the dominant fungi inside the college library.

Deteriorated samples of books were also collected. The selected samples were categorized as books of different age and colours. Isolation of fungi associated with books, was made by the routine method using czapek dox agar and potato dextrose agar medium at 5.5 pH. The isolated fungi from books in library were studied further for cellulolytic activity by weight loss test method after incubation for 10 and 15 days (Table 3) [17].

Temperature (°C) and Relative Humidity (%) were recorded in the libraries during the sampling period using a Hygrometer. (Table 1)

Table 1: Average temperature and relative humidity recorded from two Sections of library.

Months	***Temperature (°C)***				***Humidity (%)***			
	R		**S**		**R**		**S**	
	Max	**Min**	**Max**	**Min**	**Max**	**Min**	**Max**	**Min**
May	39.2	31.2	40.1	35.5	48	30	47	24
June	31.1	26.5	33.8	32.7	79	64	77	63
July	23.1	22.4	26.9	26.3	87	80	87	85
August	28.9	26.9	31.5	29.3	84	75	82	73
September	30.7	29.9	33.4	30.1	78	62	74	61
October	29.9	28.6	31.7	28.9	78	69	78	70

Abbreviations: R= Reading room, S= Store room.

RESULT AND DISCUSION

The mycoflora trapped from the air inside the college library were *Aspergillus niger, A. fumigatus, A. flavus, Alternaria alternata, Altarnaria* spp., *Penicillium* spp., *Rhizopus* spp., *Fusarium* spp., *Curvularia* spp., *Chaetomium* spp., etc. (Fig.2). Out of these some of the fungi like *Aspergillus niger, A. flavus, Penicillium* spp., *Chaetomium* spp. etc. are also isolated from the infected books in library. The occurrence of fungal species examined were more in the store room where the books are stored than that of the reading room. It may be because of wet and humid conditions in the store room which induces the occurrence of mycoflora more in store room than that of reading room. In both sections of the library, concentrations of total fungal species were contributed by the *Aspergilli* group (Table 2). Higher concentration of *Aspergilli* has been reported from many other indoor environments of study areas, like bakery, poultry shed and vegetable market. Monthly variations in the total mycofloral concentration were observed during the study [18]. The concentration of mycoflora recorded was highest in the month of August, September and October than May, June and July.

Table 2: Showing occurrence of fungal species in the two sections of a library.

Fungal types	*Library sections*	*May*	*June*	*July*	*Aug*	*Sep*	*Oct*
Aspergillus niger	R	+	+	+	+	+	+
	S	+	+	+	+	+	+
Aspergillus fumigatus	R	+	-	+	-	+	+
	S	-	+	+	+	+	+
Aspergillus flavus	R	+	-	-	+	+	-
	S	-	-	-	+	+	+
Aspergillus caespitosus	R	-	+	-	+	-	+
	S	+	-	+	+	+	+
Penicillium spp.	R	-	+	-	+	-	+
	S	+	+	+	+	-	+
Alternaria alternata	R	-	+	-	-	+	+
	S	+	-	+	-	+	-
Alternaria spp.	R	-	+	-	-	-	-
	S	-	-	+	-	+	+
Curvularia spp.	R	-	+	-	+	+	-
	S	+	-	+	+	+	+
Fusarium spp.	R	-	-	-	-	+	+
	S	-	-	-	+	+	-
Rhizopus spp.	R	+	-	-	+	+	-
	S	-	-	+	-	+	+
Chaetomium spp.	R	-	-	-	+	-	-
	S	-	-	-	-	+	-
Helminthosporium spp.	R	-	+	-	+	-	-
	S	-	-	-	-	+	+
Geotrichum spp.	R	-	+	-	+	-	-
	S	+	-	+	+	+	-
Drechslera spp.	R	-	+	-	+	-	-
	S	-	-	-	-	+	-
Unidentified Fungi	R	+	+	+	+	+	+
	S	+	+	+	+	+	+

Abbreviations: R= Reading room, S= Store room.

Cellulolytic activity of some fungi isolated from books has been caused on news paper and book paper. The activity was done by weight loss test. From the results it is clear that *Aspergillus niger* and *Chaetomium* are the most cellulose degrading fungi than the other rest. *Aspergillus, Chaetomium* and *Alternaria* affects the book paper more than the news paper. *Penicillium, Rhizopus* affects the news paper more than the book paper (Table 3) (Fig. 1). The basic deterioration of the books is then basically due to the cellulolytic activity of fungi.

Table 3: The percentage loss of substrates by some book deteriorating fungi.

Fungi tested	*NP*		*BP*	
	10 days	**15 days**	**10 days**	**15 days**
Aspergillus	18.90	21.60	30.00	35.00
Penicillium	17.50	20.00	14.20	15.90
Rhizopus	16.10	18.80	12.80	14.50
Chaetomium	29.70	32.00	30.00	34.00
Alternaria	15.80	18.40	19.90	22.70

NP = News paper, BP = Book paper.

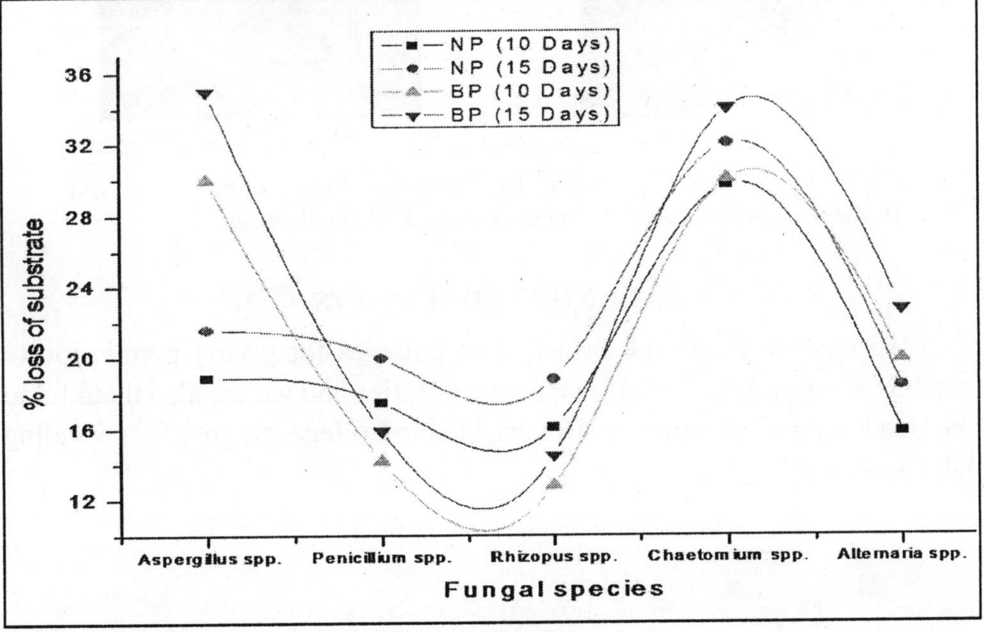

Fig.1: Percentage loss of substrates by some Fungi

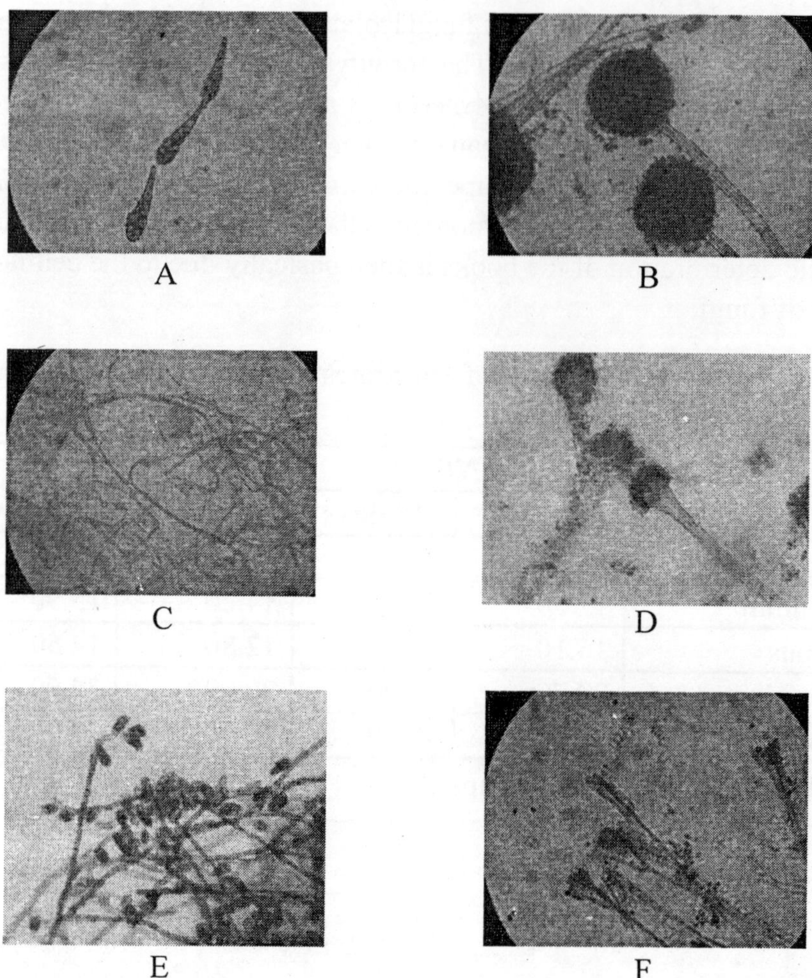

Fig.2: A. *Alternaria alternata*; **B.** *Aspergillus niger*; **C.** *Rhizopus sp.*; **D.** *Aspergillus fumigatus;* **E.** *Curvularia sp.*; **F.** *Penicillium sp.*

ACKNOWLEDGEMENTS

We wish to thank the Principal of colleges for giving permission to conduct air sampling in their respective libraries and we are also thankful to the Head of the Department of Botany, Hislop College, Nagpur for providing laboratory facilities.

REFERENCES

1. Ebner MR, Haselwandter K. Indoor and Outdoor incidence of airborne fungal allergens at low and high altitude alpine environments. Mycol Rec 1992; 96: 117-24.
2. Takahashi T.: 1997, Airborne fungal colony forming units in outdoor and indoor environments in Yokohama, Japan Mycolpathologia 139, 23-33.
3. Lacey J. The Aerobiology of conidial fungi. In: cole GT, kendrick B, editors. Biology of conidial fungi. New York: Academic Press; 1981. p. 373- 416.
4. Lehrer SB, Aukrush L, Salvaggio JE. Respiratory allergy induced by fungi. Clin chest Med 1983; 4: 23- 41.
5. Tilak, S.T. and Vishwe, D.B. 1975. Microbial content of air inside library. Biovigyanam, 1:187-190.
6. Tilak, S. T. and O. J. Chakre, 1978. Atmospheric concentration of *Cleviceps fusiforms* over Bajra fields in relation to their environment factors, IV Int. Conf.Palynol.1976-77.Lucknow, Abst.181-182.
7. Babu.M., 983. Aerobiological studies at Aurangabad.Ph.D.Thesis Marathwada University. Aurangabad.
8. Vittal, B.P.R. and Glory, A. Leela 1985'*Airborne Fungus* Spores of a Library in India', Grana, 24: 2, 129 – 132.
9. Tilak, S.T. and S.G. Pillai, 1988 Fungi in Library. An Aerobiological survey. Ind.J.Aerobiol. 1(2):92-94.
10. Singh, A., Ganguli, M. and Singh, A. B. 1995. *Fungal spores* are an important component of library air. Aerobiologia 11 1995 231- 237.
11. Sinha, A., Singh, M.K. and Kumar, R. 1998. Aerofungi-An important atmospheric biopollutant at atmosphere. Ind.J.Aerobiol.Vol. 11(1&2), 19-23.
12. Verma, K.S., and Srivastava, 2004. Airborne fungi of Poultry environment. Advances in Pollen spore research. Vol. XXII: 93-94.
13. Tiwari, K.L. (2005) Studies on aeromycoflora of dairy area at Raipur (C.G.) India. Flora and fauna, Vol. II (2) 195-196.
14 Pillai, Shanta G. and Patil, Mukundraj B, 2007 Effect of threshing on air pollution of indoor environment. Ind. J. aerobiol. 20 (I and II) 63-65.
15. Patil, Mukundraj B, Kamble, S. A. and Pillai, Shanta G. 2009 Airborne bioparticles in the industrial area. Bionanofrontire. Science Day Special issue: 85-86.

16. Lyon, F. L., Framer, C. L. and Eversmeyer. M.G. 1984. Variation of airspora in the atmosphere due to weather conditions. *Grana.* 23: 177-181.

17. Bose, B. and A. S. Yadav 1973. Some cellulolytic microfungi and their cellulose activity *in vitro*. J. Ind. Bot. Soc. 52: 218-224.

18. Adhikari, A. Investigations on aero-mycology in relation to allergy in some selected semi-rural places of West Bengal, India. 2000. PhD Dessertation, Jadavpur University, Calcutta, India.

Floral Diversity and their Conservation (2013), Editors: D.R. Khanna et al.
Published by Biotech Books.
ISBN: 978-81-7622-286-0 Pages: 267-290

28

Ethnomedicinal uses of Pteridophytes in Pachmarhi Hills, Madhya Pradesh, India

Singh Shweta*, Singh Rita* and Sahu T. R. **
*Guru Gobind Singh University, University School of Environment Management, Biosystematics Lab, 16 C Dwarka, New Delhi
**Dr. Hari Singh Gaur Vishwavidyalaya, Sagar (M.P.)

The pteridophytes are the second largest component of the world flora. These plants are widely used by the local tribes and the plants materials are sold in the local market of Pachmarhi hills, Madhya Pradesh, India. The present study revealed that 48 species of pteridophytes belonging to 27 families and 34 genera are traditionally used by tribal communities of Gond, Korku, Bharia, Bhil and Mabasi of the region in the treatment of various diseases which contribute to about 36% of total pteridophytic diversity of the area. The plant parts, *viz.* leaves, roots, rhizomes and fronds were used in raw or cooked forms for the treatment of malaria, gonorrhea, jaundice, leprosy, rheumatism etc. Ethnomedicinal information was gathered through questionnaire from the tribal and non-tribal people of Pachmarhi hills. Data set have been recorded along with Latin names, vernacular names, families, the locality of occurrence, their habit and habitats along with the parts utilized for medicinal uses. Our study concluded that, the wealth of indigenous ethnomedicinal knowledge of teridophytes may also points to a great potential for research in the discovery of new drugs to fight diseases and other new uses.

Key words: Ethnomedicinal, Pteridophytes, Pachmarhi Hills, Madhya Pradesh, India

INTRODUCTION

For centuries, human beings have been utilizing plant genetic resources for food, medicine and other cultural purposes. The local plant resources are the principal source of medicine and are used by the traditional herbal healers. Hundreds of plants growing in forests are used as source of medicines throughout the world. Some of the plants have pharmacological properties while the others are used in indigenous medicine. Most of these plants has occupied an important place in the past and shall continue in the coming days in traditional as well as in modern medicine system. The pteridophytes have a long geographical history on our planet. These fascinating groups of plants are always attracting the botanist and naturalist all over the globe not only because of its beautiful and unique foliage but also because of their medicinal properties. They are more abundant in hilly and mountainous regions at high altitude. In comparison to higher plants they have found very little applications in medicine and are still neglected. The numbers of contributions about the taxonomy, ecology and distribution of pteridophytes have been published time to time in various ways but enough attention has not been paid towards their useful aspects.

The Pachmarhi widely known as ***"Satpura ki Rani"*** (Queen of Satpura), situated at a height of 1000 m in a valley of the Satpura Range in Hoshangabad district, M.P. It laying at about 22° 11' to 22° 56' N latitude and 77° 47' to 78° 52' E longitude and the other neighboring hilly tracts falling in Madhya Pradesh, have attracted the attention of botanical workers since the ancient age. The area of the present study covers Pachmarhi proper and the adjoining hills. The total area of 417.78 sq. km covered by the plateau practically 50-60% is dominated by forest. The altitude varies from 320 to 1385m. Rising to altitudes of more than 1300 meters above sea level, these mountains have rich forest covers, excellent bio-diversity, a large variety of wildlife species and, above all, hundreds of perennial fresh-water streams rushing down their slopes. This combination of mountains and water here has carved hundreds of caves, cave shelters and nature-made tunnels that became home to one of the earliest human civilizations. The average annual rainfall is about 200–250 cm. With cascading waterfalls, ravines and gorges sculpted into the red sandstone earth, and dense evergreen sal and teak forests interspersed with wild bamboo groves, it is a veritable paradise for botanists and geologists. Pachmarhi is also an archaeological treasure-house, with rock paintings in cave shelters, some of which are estimated to be 10,000 years old.

According to Champion (1968) Pachmarhi forests are classified as Central Indian Sub – Tropical Evergreen Forests of the Southern Sub Tropical Wet Hill Forests. The flora of Pachmarhi consists of a mixture of temperate and tropical elements. The forest is extremely dense, covering nearly 80% of the total land area. Perennial streams and dark shady gorges encourage the growth of moisture loving species such as pteridophytes, orchids and rare herbs. The area is generally perceived as a tribal dominated area. Many ethic groups and folklore *viz.* Korku, Gond, Bharia, Bhil, Mauria, Maria, Paria, Bhatara and Baigas inhabit the area, so there is an ample scope to explore ethnobotanical and traditional knowledge. They are living in very remote areas and have largely been untouched by the mainstream, which are inaccessible and in isolation. Therefore, tribals are largely dependent upon the use of folk medicines to cure diseases and various ailments form which they suffer. In turn they protect the forests, conserve its diversity, and also enrich fertility with various cultural activities. They have their own life style and have retained more of their traditional culture. They also have a special skill that has been passed down every generation.

In the lack of proper education, excessive collection, rapid industrialization and biotic interference the number of these useful species of pteridophytes are likely to be decreasing continuously, therefore it is urgent need to formulate conservation strategies to save the medicinally rich pteridophytic plant diversity occurring in the area. The aims and objective of this investigation is to collect the information based on medicinal pteridophytes commonly used by the ethnic communities and to document these ethnobotanical as well as ethnomedicinal information at one place which are prevalent in the region of study and may be lost if not properly documented.

METHOLOGY

Extensive survey cum collection practices was made in the rural and tribal pockets of Pachmarhi and the adjoining areas. The information regarding the uses of pteridophytic species was gathered from the local villagers as well as some tribal people. The data presented are based on the first hand information collected during the period in months of October and November 2010. For the study of all kinds of vegetation field tours were conducted to different localities around the Pachmarhi places like Numbingly Mahadeo, Chauragarh, Sangam Points, Vanushree Vihar, Dhupgarh, Jamboodweep, Bee Fall, Dutches Fall, Rorighat, Bori, Beldhandar, Panarpani, Bawadi, Denwa forest, Priyaarshni point, Gupt Mahadev, Pansy Pool, Piccadilly circus, Rajendra Giri, Lanjee Giri, Patharchatta, Pachmari lake, forest near Madai, Tawa Dam, Crumps crag, Lady Robertson's view, Colletin Crag, Mount Rosa, Naagdwari and Matkuli during various seasons. This information collected was verified by cross-checking with tribal living herbal medicine practitioners of various ethnic groups though interviews, discussions, personal contacts and keen observations (Plate 3: 17). The herbarium specimens have been deposited in the Botanical Survey of India, Central circle Allahabad (BSA).

RESULT

The present study revealed that 48 species of pteridophytes belonging to 27 families and 34 genera are traditionally used by tribal communities of the region in the treatment of various diseases which contribute to about 36% of total pteridophytic diversity (134 species) of the area. The present paper provides botanical names, specific characters, ecology, vernacular names, families, plant parts used for various ailments and ethnomedicinal values of the species recorded from the area. This paper also compares the present information with that of earlier ethnobotanical observations recorded.

ENUMERATION

1. ***Actiniopteris radicata*** (Sw.) Link (Family: Pteridaceae)
Local name: *Morpankhi, Mayur-shikha*

Small xerophytic plant, lamina fan-like with dichotomously segments and rooting in rocky soil long roadside usually in lime rich/alluvial soil. (Plate 1:1)

Uses: Plant used as styptic and anthelmentic. The fresh fronds chewed for sore throat and the decoction used in cure of dandruff. (cf. Dymock, 1890; Dixit, 1975a; Singh *et al.*, 2003; Dixit & Singh; 2004; Singh *et al.*, 2005).

2. *Adiantum capillus-veneris* L. (Family: Pteridaceae)

Local name: *Hansraj;* English name: *Southern maidenhair fern, Venus-hair fern*

Terrestrial plants, fronds bipinnate and rooting in moist soil cover over rocks along flowing streams in shady situations, forming pure populations. (Plate 1:2)

Uses: Plant used as diuretic, emmenagogue, febrifuge expectorant, demulcent and astringent. The decoction of leaves along with tea is used in abnormal stoppage of menses. The steam of fresh fronds is very effective in cure of smallpox and decoction used in cure of ricketts in children. (cf. Quisumbing, 1955; Scully, 1970; Dixit, 1975b; Rama Shankar *et al.*, 1992; Uddin *et al.*, 1998; Vasudeva, 1999c; Singh *et al.*, 2003; Dixit & Singh; 2004; Singh *et al.*, 2005).

3. *Adiantum incisum* Forssk. (Family: Pteridaceae)

Local name: *Mayur-pambi***;** English name: *Walking fern*

Terrestrial walk fern and growing in the forest floor along the edges of forests, forming patches. Fronds curl up in dry weather. (Plate 1:3)

Uses: Plant used in cure of skin diseases, fever, cough and diabetes. The fresh fronds chewed for curing mouth blisters. The decoction is used in problem of bronchitis. The powder of leaves is mixed with butter is used for controlling the internal burning of the body. (cf. Pande *et al.*, 1994; Uddin *et al.*, 1998; Vasudeva, 1999c; Singh *et al.*, 2003; Dixit & Singh; 2004; Singh *et al.*, 2005).

4. *Adiantum philippense* L. (Family: Pteridaceae)

Local name: *Kali-Jhant, Hansraj*; English name: *Maiden hair*

Terrestrial fern and growing on rock boulders along streams in hilly regions or under heavy rock boulders.

Uses: Fronds used in fever, asthama, bronchitis, dysentery, epileptic fits, leprosy, ulcers and erysipelas. Rhizomes give against dog-bite and snake-bite as an antidote. It is also used in cure of glandular swelling. The extract of leaves is taken orally and paste of leaves is applied on the lower portion of stomach for clear and early release of urine. (cf. Dixit, 1975a; Rama Shankar *et al.*, 1992; Uddin *et al.,* 1998; Vasudeva, 1999c; Manickam, 1999; Singh *et al.,* 2003; Dixit & Singh; 2004; Singh *et al.,* 2005).

5. ***Alsophila gigantea*** **Wall. ex Hook.** (Family: Cyatheaceae)
English name: *Tree fern*
Tree fern, 2-3 m high and growing on hilly tracts along flowing streams in humid situations.
Uses: Extract of grinded rhizome and young petiole is given in snake bite. (cf. Rama Shankar *et al.,* 1992; Singh *et al.,* 2003; Dixit & Singh; 2004; Singh *et al.,* 2005).

6. ***Alsophila spinulosa*** **(Wall. ex Hook.) Tryon** (Family: Cyatheaceae)
English name: *Tree fern*
Large tree fern, spines throughout, growing on hilly tracts along flowing streams in humid situations. (Plate 1:4)
Uses: Leaves are locally applied on wounds. (cf. Manickam, 1999; Singh *et al.,* 2003; Dixit & Singh; 2004; Singh *et al.,* 2005).

7. ***Angiopteris evecta*** **(Frost.) Hoffm.** (Family: Angiopteridaceae)
Local Name: *Ghora tap*
Large fern, fronds tufted lamina bipinnate and found in humid slopes near water fall.
Uses: The fleshy rhizome is eaten by tribals during scarcity. The pith and rhizome boiled with mustard oil is used as massage for rib pain and fronds are bounded to fractured limb. Whole plant extraction used as tonic. Rhizome are used in preparation of intoxicating drink called as 'ruchshi'. (cf. Dixit *et al.,* 1978; Jain, 1991; Dixit, 1993; *Uddin et al.,*1998; Singh *et al.,* 2003; Dixit & Singh; 2004; Singh *et al.,* 2005).

8. ***Asplenium falcatum*** **Lam.** (Family; Aspleniaceae)
A small tufted fern and growing as lithophytes along streams.
Uses: Fronds used in malaria, jaundice, enlargement of spleen, urine trouble and calculus.
(cf. Chopara *et al.*,1956; Uddin *et al.,*1998; Dixit & Sinha, 2001; Singh *et al.,* 2003; Dixit & Singh; 2004; Singh *et al.,* 2005).

9. ***Asplenium trichomanes*** **L.** (Family; Aspleniaceae)
English name: *Common spleen wart*

A small terrestrial fern, covered with dark scale, fronds tufted and growing on rocks deep into the forest.

Uses: Plant bitter, laxative, expectorant used in pulmonary disease and in abscess of uterus. Rhizome used as anthelminthic and for hypochondriac infections. Whole plant extract is used against enlargement of spleen. (cf. Bir *et al.,* 1983; Asolkar *et al.,* 1992; Vasudeva, 1999c; Kirn & Kapahi, 2001; Singh *et al.,* 2003; Dixit & Singh, 2004; Singh *et al.,* 2005).

10. *Athyrium filix-femina* (L.) Roth. (Family: Woodsiaceae)
English name: *Common Lady Fern*
Terrestrial plants, clothing with brown scales, fronds tufted, growing on heavy rock boulders in moist shady situations along nalas, streams etc. (Plate1:5)

Uses: Rhizome vermifugal and used as anthelminthic. (cf. Vasudeva, 1999c; Chopara *et al.,*1956; Singh *et al.,* 2003; Dixit & Singh, 2004; Singh *et al.,* 2005).

11. *Azolla pinnata* R. Br. (Family: Azollaceae)
English name: *Mosquito fern, Duckweed fern, Fairy moss*
A small floating aquatic fern and elder leaves dull red in colour. Common in cultivated fields, pond, and in low lying areas. (Plate 1:6)

Uses: The plant is used as diuretic and also has antibacterial properties. (cf. Uddin *et al.,*1998; Singh *et al.,* 2003; Dixit & Singh, 2004; Singh *et al.,* 2005).

12. *Blechnum orientale* L. (Family: Blechnaceae)
English name: *Centipede Fern*
A terrestrial fern with frond large. Plants growing along the ponds and streams in the forest edges. (Plate 2:7)

Uses: Whole plant used as medicine for diarrhea, eradication of worms and disorders related to gallbladder. Plants used as poultice for boils. It is used in impotency in men, urinary disorders, and in cure of sanipat (Delirium). It is used as depurative and appetitive. (cf. Quisumbing, 1951; Dixit, 1975a; Kaushik & Dhiman, 1995; Uddin et *al.,*1998;Vasudeva, 1999c; Singh *et al.,* 2003; Dixit & Singh, 2004; Singh *et al.,* 2005).

13. *Ceratopteris thalictroides* (L.) Ad. Brongn. (Family: Pteridaceae)
Local name: *Pani ka karela,* English name: *Water fern*

An aquatic dimorphic fern and growing along the fresh water pounds, shallow ditches and mud in open places.
Uses: Whole plant used as tonic and styptic. Fronds are used as poultice in skin complains.
(cf. Dixit, 1975a; Uddin *et al.,*1998; Vasudeva, 1999c; Singh *et al.,* 2003; Dixit & Singh, 2004; Singh *et al.,* 2005).

14. *Cheilanthes albomarginata* C.B.Clarke (Family: Pteridaceae)
Local name: *Bhoot kesari, Nanha*
Small xerophytic terrestrial fern tufted fronds with yellow-white powder in lower side and rooting in rock crevices or along newly built forest in sunny situations.
Uses: Whole plants used against tuberculosis and as tonic. It is also known to have antibacterial properties. (cf. Vyas & Sharma, 1988; Uddin *et al.,*1998;Vasudeva, 1999c; Singh *et al.,* 2003; Dixit & Singh, 2004; Singh *et al.,* 2005).

15. *Cheilanthes bicolor* (Roxb.) Fras.-Jenk. (Family: Pteridaceae)
Small terrestrial herb rooting in rock crevices or along the road side near forest in sunny situations.
Uses: Whole plants used in seasonal fever and cold. (cf. Asolkar *et al.,* 1992; Vasudeva, 1999c).

16. *Cheilanthes farinosa* (Forssk.) Kaulf. (Family: Pteridaceae)
Local name: *Chooti Brahmi.*
Small xerophytic fern rooting in rock crevices or under heavy rock boulders, along newly constructed roads, in lime rich soil. (Plate 2:8)
Uses: Extract of Rhizome is useful to the unconscious patients suffering from epilepsies. Paste of root applied in cure of eczema and wounds. Decoction of root used in stomachache. Fronds used to treat menstrual disorders (cf. Rama Shankar *et al*., 1992; Singh *et al.,* 2003; Dixit & Singh, 2004; Singh *et al.,* 2005).

17. *Cheilanthes tenuifolia* (Burm.f.) Sw. (Family: Pteridaceae)
Local name: *Veli-chhoti, Dodhari.*
A small, erect xerophytic fern and grows in forest floor along roadsides, rooting in rock crevices in open sunny situations. (Plate 2:9)
Uses: There is a superstition among the tribal that preparation made from roots is given for sickness attributed to "evil eye" or "witch craft". It is used

as tonic and the paste of roots with 'karanj oil' is applied on wound for a week. (cf. Rama Shankar *et al.,* 1992; Uddin *et al.,*1998; Vasudeva, 1999c; Manickam, 1999; Singh *et al.,* 2003; Dixit & Singh, 2004; Singh *et al.,* 2005).

18. *Dicranopteris linearis* (Burm.f.) Underw. (Family: Gleicheniaceae)
English name: '*Thicket fern*
Terrestrial, widely spreading and sub scandent fern rooting on rock boulders with little humus or in rocky alluvial soil, along flowing streams, in exposed places forming thickets.
Uses: Fronds are useful in cure of asthma. It also has antibacterial properties. Young fronds with cow's milk given in woman's sterility. Rhizome anthelminthic. Spores used as favorite remedy for diarrhoea in children. (cf Rama Shankar *et al*., 1992; Kaushik & Dhiman,1995; Uddin *et al*.,1998; Vasudeva, 1999c; ; Singh *et al.,* 2003; Dixit & Singh, 2004; Singh *et al.,* 2005).

19. *Diplazium esculentum* (Retz.) Sw. (Family: Woodsiaceae)
English name: *Vegetable fern*
Large fern growing along streams in open places, in the edge of forest.
Uses: Young tips of fronds are used as tonic for health. Decoction of rhizome and young leaves are useful for haemoptysis and constipation. (cf. Quisumbing, 1951; Rama Shankar *et al.,* 1992; Kaushik & Dhiman, 1995; Singh *et al.,* 2003; Dixit & Singh, 2004; Singh *et al.,* 2005).

20. *Drynaria quercifolia* (L.) J. Smith (Family: Drynariaceae)
Local name: *Aswakatri*; English name: *Oak-leaf fern*
Plants growing as epiphytes or lithophytes among rock boulders along streams in low-land areas. Lamina green when young turning dark-brown at age.
Uses: Whole plant used in hectic fever, dyspepsia, cough, dyspepsia, poulticing swelling and typhoid fever. Rhizome anthelminthic, expectorant, pectoral and astringent. The fluid extract of fronds shows antibiotic properties. (cf. Kirtikar and Basu,1918; Chopara *et al.,* 1956.;Dixit, 1975; Singh *et al.,* 2003; Dixit & Singh, 2004; Singh *et al.,* 2005).

21. *Dryopteris cochleata* (Buch. Ham. ex D.Don) C.Chr.
(Family: Dryopteridaceae)
Local name: *Jatashankari*

Tufted large, herbaceous fern, grows in rich laterite soil along the streams and nalas in the sunny situations.

Uses: Juice extract of leaves is given in epilepsy and leprosy. Whole plant paste used against snakebite. Rhizome is used in swellings and pain and has antifungal properties. (cf Kaushik & Dhiman, 1995;; Singh *et al.*, 2003; Dixit & Singh, 2004; Singh *et al.,* 2005).

22. *Dryopteris sparsa* (Buch.-Ham. *ex* D. Don) O. Ktze.

(Family: Dryopteridaceae)

Terrestrial plant, herbaceous, growing along the forest floor along streams and nalas in sunny situations.

Uses: Rhizome have anthelminthic properties.(cf. Manickam, 1999; Singh *et al.,* 2003; Dixit & Singh, 2004; Singh *et al.,* 2005).

23. *Equisetum diffusum* D.Don (Family: Equisetaceae)

English Name: *Horse tail*

Small terrestrial to sub aquatic perennial plants, occur in the edge of forest in water logged shady places.

Uses: Whole plants are useful in acidity and dropsy. (cf. Manickam, 1999; Singh *et al.,* 2003; Dixit & Singh, 2004; Singh *et al.,* 2005).

24. *Equisetum ramossissimum* Desf. subsp. *debile* (Roxb. ex Vauch.) Hauke (Family: Equisetaceae)

Local name: *Jod ka tod*; English name: *Fox tailed Asparagus*

Large terrestrial to sub aquatic perennial plants, growing in shady or open moist / wet places preferring sandy-alluvial soil at lower elevations, trailing through bushes.

Uses: The plant is useful in cure of gonorrhoea and treatment of fracture. (cf. Bir *et at.,* 1983; Singh *et al.,* 2003; Dixit & Singh, 2004; Singh *et al.,* 2005).

25. *Hypodematium crenatum* (Frossk.) Kuhn v. Desk.

(Family; Hypodematiaceae)

Terrestrial fern, herbaceous and growing among rocks along stream.

Uses: A whole plant is used for gynecological disorders and also having antibacterial properties. Scales of this fern are used against the effect of "Witch craft or evil eye". Leaves used to facilitate conception in women. (cf. Vyas and Sharma, 1988; Singh *et al.,* 2003; Dixit & Singh, 2004; Singh *et al.,* 2005).

26. *Isoetes coromandelina* L.f. (Family : Isoetaceae)

English name: *Coromandel quillwort*

Fresh water aquatic or amphibious plant. One of the common species occurring along the edges of tanks (artificial ponds), streams, paddy fields, and shallow ponds, growing during the rainy season.

Uses: Orally used in spleen disorders and liver complains (cf. cf. Dixit *et al.*., 1978; Singh *et al.,* 2003; Dixit & Singh, 2004; Singh *et al.,* 2005).

27. *Leucostegia immersa* (Wall. ex Hook.) Presl

(Family: Davalliaceae)

Herbaceous fern and hanging downwards from the rocks or epiphytic at the base of the tree trunks. (Plate 2: 10)

Uses: The leaves are useful in curing constipation. (cf. Dixit *et al,* 1978; Singh *et al.* 2003; Dixit & Singh, 2004; Singh *et al.,* 2005).

28. *Lygodium flexuosum* (L.) Sw. (Family: Lygodiaceae)

English name: *Climbing fern*

Climbing ferns growing along the edges of forest, roadsides and climbing on bushes and trees or trailing on the ground, rooting in gravelly, sandy soil near streams.

Uses: Fronds boiled with mustard oil useful as local application to carbuncles externally in rheumatism, sprains, scabies, ulcers, eczema and cut wounds. Rhizome cures gonorrheoa and paste applied on the piles. Spores cure high fever. An infusion of plant is used in menorrhagia and also having antibacterial properties. (cf. Chopara *et al.,* 1956; Singh *et al.,* 2003; Dixit & Singh, 2004; Singh *et al.,* 2005).

29. *Marsilea minuta* L. (Family: Marsileaceae)

Local name: *Chaupatira*; English name: *Dwarf water clover*

Small aquatic fern, growing in plains during rainy season and along the edges of water ponds, ditches, paddy fields etc.

Uses: Decoction of leaves mixed with ginger given in bronchitis, spastic condition of leg, cough, sedative and insomnia. Fresh young leaves are in form of juice is effective in urinary disorder. Leaves extract with sugar or fishes are useful in migraine. Drops of juice are very effective eye disease. Also having antibacterial properties. (cf. Roy & Gupta, 1965; Jha & Varma, 1996; Singh *et al.,* 2003; Dixit & Singh, 2004; Singh *et al.,* 2005).

30. ***Microlepia speluncae*** **(L.) Moore** (Family: Denstaedtiaceae)
English name: *Limp leaf Fern, Lace Fern*
Large often straggling fern, rhizome wide-creeping, fronds distant, 50-100 cm tall, pubescent-hispid, bipinnate and occurs in the edge of forest under shady situations.
Uses: Leaves are effective in curing of high fever. (cf. Dixit & Sinha 2001; Singh *et al.,* 2003; Dixit & Singh, 2004; Singh *et al.,* 2005).

31. ***Nephrolepis cordifolia*** **(L.) C. Presl** (Family: Oleandraceae)
English name: *Fish bone fern*
A tufted, wiry fern, growing along streams and nalas in sunny situations.
Uses: Rhizome and tubers are used in intestinal and kidney disorders. Extract of rhizome is used in permanent curing of sterility in women. Decoction used for cough. Rhizome stock is applied locally in skin disease (cf. Rama Shankar *et al..*, 1992; Singh *et al..*, 2003; Dixit & Singh, 2004; Singh *et al..*, 2005).

32. ***Ophioglossum costatum*** **R.Br.** (Family: Ophioglossaceae)
Local name: *Sheambli*
Terrestrial small erect fern, growing among grasses in open places or under shade of tree.
Uses: Dried tubers are crushed in the forms of power mixed with mustard oil used in skin disease. (cf. Dadhich & Sharma 2002; Singh *et al.,* 2003; Dixit & Singh, 2004; Singh *et al.,* 2005).

33. ***Ophioglossum reticulatum*** **L.** (Family: Ophioglossaceae)
Local name: *Vanpalak, Gondi.*
Terrestrial plants, growing in the forest floor amidst grasses/mosses, etc. or in moist alluvial, sandy soil or heavy rock boulders along flowing streams with little humus. This seems to be one of the most common and polymorphic species since a number of intermediate forms with various interlinked shapes of tropophyll are observed. (Plate 2:11)
Uses: Fronds are used in preparation of tonic. It is useful in relief of headache. Paste of fresh leaves and tubers applied on boils and also used as cooling agent. (cf. Singh *et al.,* 1989; Vasudeva, 1999c; Singh *et al.,* 2003; Dixit & Singh, 2004; Singh *et al.,* 2005).

34. ***Osmunda regalis*** **L.** (Family: Osmundaceae)
English name: *Royal Fern*

Terrestrial plants abundantly grow in exposed marshy places, in humid slope near water fall, bank of streams and evergreen forest along stream side on sandy loam soil.

Uses: Plants used as tonic and styptic. Young fronds cure toothache, rheumatism, rickets and intestinal gripping. Fronds form a constituent of diuretic drinks given for swelling. Rhizome used as abortifacient and roots are stimulant, having antibacterial properties and useful in cure of dysentery and muscular debility. Tender shoots used in balms and healing plasters (cf. Chopara *et al.,* 1956; Vasudeva, 1999c; Singh *et al.,* 2003; Dixit & Singh, 2004; Singh *et al.,* 2005).

35. *Palhinhaea cernua* (L.) Franco et Vasc. (Family: Lycopodiaceae)
Local name: *Rumi-Jhumi*

Terrestrial fern. Trailing on ground among grasses in open sunny places near water current or streamside. Abundant.

Uses: Leaves are effective in skin eruptions and beri-beri. (cf. Dixit & Sinha, 2001; Singh *et al.,* 2003; Dixit & Singh, 2004; Singh *et al.,* 2005).

36. *Pteris biaurita* L. (Family: Pteridaceae)

Terrestrial large fern growing along the edges of forests, generally in open places along stream sides.

Uses: The paste of stipe and stem is useful in wounds. (cf. Manickam, 1999; Singh *et al.,* 2003; Dixit & Singh, 2004; Singh *et al.,* 2005).

37. *Pteris vittata* L. (Family: Pteridaceae)
English name: *The Chinese brake fern*

Densely tufted ferns growing in shady or moist situations near water fall or in the city areas on dilapidated moist walls. (Plate 2:12)

Uses: Rhizome used as demulcent. It used in glandular swelling. (cf. Dadhich & Sharma, 2002; Singh *et al.,* 2003; Dixit & Singh, 2004; Singh *et al.,* 2005).

38. *Pteridium aquilinum* (L.) Kuhn. ex Decken (Family: Pteridiaceae)
English name: '*Bracken fern*'

Terrestrial tufted fern growing in hilly regions along streams-sides and in shady situations.

Uses: Rhizome has astringent and anthelminthic properties. The decoction of rhizome and fronds is given in chronic disorders arising from obstructions of viscera and spleen. (cf. Dixit, 1975a; Manickam, 1999; ; Singh *et al.*, 2003; Dixit & Singh, 2004; Singh *et al.*, 2005).

39. *Psilotum nudum* (L.) P. Beauv. (Family: Psilotaceae)
Local name: *Bhulbhari*
A rootless perennial herb growing as epiphytes or lithophytes; aerial shoots erect or hanging downwards along streams with perennial source of flowing water. (Plate 3:13)
Uses: Whole plant used as purgative. The oily spores are given to infants to arrest diarrhoea; the juice of the herb showed antibacterial activity. (cf. Uddin *et al.*.,1998; Singh *et al.*, 2003; Dixit & Singh, 2004; Singh *et al.*, 2005).

40. *Pyrrosia adnascens* (Sw.) Ching (Family: Polypodiaceae)
Local Name: *Bormondi*
Epiphytes or lithophytes in moist shady situations along flowing streams. Scarce. The fronds are curled up in absence of moisture. It is a drought resistance species.
Uses: Fronds used in skin burn and in dysentery. (cf. Manickam, 1999; Singh *et al.*, 2003; Dixit & Singh, 2004; Singh *et al.*, 2005)

41. *Selaginella bryopteris* (L.) Baker (Family: Selaginellaceae)
Local name: *Sangivani*
Xerophytic plants growing on heavy rock boulders forming thick, green carpet during rainy season. Leaves curled up in dry weather but retain original colour and shape if dipped upside down in water for sometime. (Plate 3:14)
Uses: Whole plans diuretic and gonorrhea. Dried plants with Tobacco smoked for hallucination. Leaves with sugar are taken in stomachache, inflammation of urinary tract and in some venereal disease. A popular strength tonic amongst local people. Effective in spermatorrhoea and leucorrhoea. (cf. Shah & Singh, 1990; Manickam, 1999; Vasudeva, 1999c; Singh *et al.*, 2003; Dixit & Singh, 2004; Singh *et al.*, 2005).

42. *Selaginella ciliaris* (Retz.) Spring (Family: Selaginellaceae)
Small plants, ciliated at base and rooting in rock crevices along streams with permanent source of trickling water. It is one of the common species.

Uses: Plant extract is used in stopping bleeding, coughs, prolapse of rectum, gravel and amenorrhoea. (cf. Singh *et .al,* 2001; Singh *et al.,* 2003; Dixit & Singh, 2004; Singh *et al.,* 2005).

43. *Sphenomeris chinensis* (L.) Maxon (Family: Lindsaeaceae)
Terrestrial ferns grow in hilly regions along stream-sides but sunny situations rooting in lime rich soil.
Uses: Fronds paste used internally for chronic enteritis. Also having antibacterial properties. (cf. Caius, 1935; Uddin *et al..,*1998; Singh *et al.,* 2003; Dixit & Singh, 2004; Singh *et al.,* 2005).

44. *Tectaria coadunata* (Wall. ex Hook. et Grev.) C.Chr.
(Family: Dryopteridaceae)
Terrestrial membranaceous ferns growing among rock boulders along flowing streams.
Uses: Thc dccoction of rhizomc is given in stomachache (anthelmintic) of children. Also used in cure of Asthma, bronchitis and getting relief from insects bites (cf. Pande *et al.,*1994; ; Singh *et al.,* 2003; Dixit & Singh, 2004; Singh *et al.,* 2005).

45.*Thelypteris arida* (D.Don) C.V. Morton (Family: Thelypteridaceae)
An erect terrestrial fern growing along stream near flowing water.
Uses: Rhizome decoction is useful in rheumatism and veterinary larval infections. (cf. Manickam, 1999; Singh *et al.,* 2003; Dixit & Singh, 2004; Singh *et al.,* 2005)

46. *Thelypteris parasitica* (L.) Tardieu (Family: Thelypteridaceae)
Terrestrial fern growing along ravines in the edge of flowing water. (Plate 3:15)
Uses: Fronds are useful against rheumatism. (cf. Manickam, 1999; Singh *et al.,* 2003; Dixit & Singh, 2004; Singh *et al.,* 2005).

47. *Thelypteris prolifera* (Retz.) C.F. Reed (Family: Thelypteridaceae)
A terrestrial spreading fern, indefinite length spreading through buds on rachis giving new plant freely and trailing through grasses and bushes in the edge of the forest under wet-moist places.
Uses: Plant having antibacterial properties. (cf. Uddin *et al.,*1998; Singh *et al.,* 2003; Dixit & Singh; 2004; Singh *et al.,* 2005).

48. *Trigonospora ciliata* (Wall. ex Benth.) Holtt.
(Family: Thelypteridaceae)
Terrestrial plants grow on rocks near water stream in moist situations. (Plate 3:16)
Uses: Arial part and rhizome used in making medicine for strength. (cf. Kaushik & Dhiman, 1995; Singh *et al.,* 2003; Dixit & Singh, 2004; Singh *et al.,* 2005).

DISCUSSION AND CONCLUSION

The present work is the result of intensive systematic traditional and ethnomedicinal observation conducted in Pachmarhi (M.P.). All the forty eight species of pteridophyes of traditional and ethnomedicinal interest are recorded after critical screening with the available literature. These are recommended for further phytochemical/pharmacological investigation and nutritional analysis, which might result in the discovery of new drug molecules for human welfare. These plants are extracting widely along with the rhizome resulting rarity of species from the forest area of the Pachmarhi. Therefore it is essential to conserve these species before extinction by way of their cultivation and stopping wide extraction from the forest area. Traditional and folklore medicine handed down from generation to generation is rich in domestic recipes and communal practice. It is interesting to note that rural communities of tribals still dependent on herbal medicines and even today they hesitate to use modern medicines. They uses (a) Decoction of rhizome, roots, stipe, stem, root-bark, spores and fronds; (b) Juice of fronds, stipe and stem; (c) Paste prepared from rhizome, roots and fresh fronds; (d) Spores are powdered for oral consumption and (e) Oil extracted from different plant parts. In India about 1200 species of pteridophytes are distributed throughout out of which numerous species are traditionally used by the local tribal. They may be lost if their traditional, indigenous and ethnomedicinal knowledge not properly documented. It is hoped that the workers in India may also initiate work on this neglected group of plant (pteridophtyes).

ACKNOWLEDGEMENTS

Grateful thanks are due to Dr. Prodyut Bhattacharya, Professor and Dean, University School of Environment Management, Guru Gobind Singh Indraprastha University, New Delhi for the encouragement and facilities. We are also thankful to UGC, New Delhi for providing financial grant under PDF Scheme.

PLATES 1:

1. *Actiniopteris radicata* (Sw.) Link

2 *Adiantum capillus-veneris* L.

3. *Adiantum incisum* Forssk.

4. *Alsophila spinulosa* (Wall. ex Hook.) Tryon

5. *Athyrium filix-femina* (L.) Roth.

6. *Azolla pinnata* R. Br.

PLATES 2:

7. *Blechnum orientale* L.

8. *Cheilanthes tenuifolia* (Burm.f.) Sw.

9. *Cheilanthes farinosa* (Forssk.) Kaulf.

10. *Leucostegia immersa* (Wall. ex Hook.) Presl

11. *Ophioglossum reticulatum* L.

12. *Pteris vittata* L.

PLATES 3:

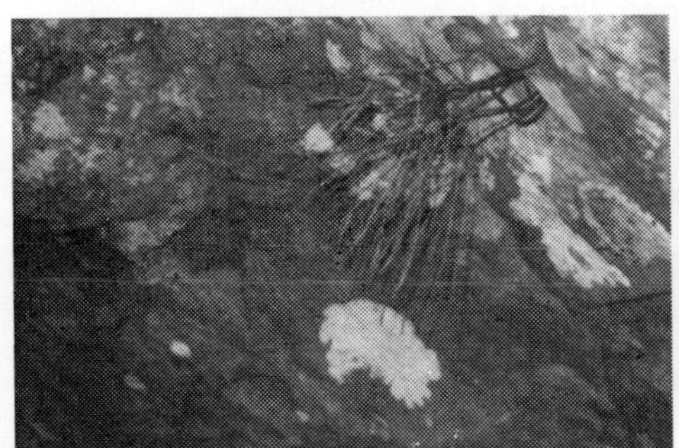

13. *Psilotum nudum* (L.) P. Beauv.

14. *Selaginella bryopteris* (L.) Baker

15. *Thelypteris parasitica* (L.) Tardieu **Holtt.**

16. *Trigonospora ciliata* (Wall. ex Benth.)

17. Author interviewing the local people about the utility of ethnomedicinal pteridophytes in Pachmarhi (M.P.)

REFERENCES

1. Asolkar, L.V., Kakkar, K.K. and Chakae, O.J. 1992. Glossary of Indian Medicinal plants with active principals. Pt. 1.CSIR. New Delhi.
2. Anand, R K & Srivastava, R B 1994. Ethonopharmacological study of *Adiantum lunulatum* Burm. f. Indian Fern. J. 11:137-14.
3. Ambasta, S.P. (Ed.)1986. The useful plants of India CSIR. New Delhi.
4. Bhardwaja T. N & Garg, A. 1984. The antifertility effect of an Australian species of theaquatic fern Marsilia minuta L. Indian Fern J.1: 75-82.
5. Bir, S. S.1988. Evolutionary trends in the Pteridophytic Flore of India presidential address (Section of Botany). 75th Sess. Ind. Sci. Congr. Pune. 1-56.
6. Bouquet, A 1974. Plantes Medicinales de la cote d' Ivoire Oratom. Paris.
7. Brahaman, M. A & Saxena, H. O 1990. Ethonobotany of Gandhamardan Hills—Some note worthy Folk Medicinal uses Ethnobotany 2 : 71-79.
8. Burkill, I. H 1966. A Dictionary of the economic products of the Malaya Peninsula, (Gov't of Malaysia and Singapore).
9. Chopra, R. N, Nayar S L & Chopara, L C 1956. Glossary of Indian Medicinal Plants. (CSIR New Delhi).
10. Caius J. F 1935. The medicinal & poisonous ferns of India J. Bombay Nat. Hist. Soc. 38: 341-361.
11. Croft, Jim. 1982. Ferns and Man in New Guinea. Centre for plants biodiversity research Papua, (New Guinea Botany Society).
12. Dadhichi, L K & Sharma, A. P. 2002. Biodiversity strategies for conservation (Dr. S. K. Aggarwal Comm. Vol.) A P H Publishing Corporation New Delhi. 223-235.
13. Das, A. K. 1997. Less known uses of plants among the aids of Arunachal Pradesh Ethnobotany 9: 90-93.
14. Dixit R D 1974. Fern—a much neglected group of medicinal plants –I J. Res. Indian Med. 9 (4): 59-68.
15. Dixit R D 1975a. Fern—a much neglected group of medicinal plants –I J. Res. Indian Med. 10(2): 74-90.
16. Dixit R D 1975b. Fern—a much neglected group of medicinal plants –I J. Res. Indian Med. 10(2): 68-76.
17. Dixit, R D., Das, Anjali & Kar, B. D.1978. Studies on Ethnobotany-III. On some less known edible

18. Ecomomic and Medicinal ferns of Darjeeling District.W. B.Nagarjun 21:10. 1- 4.
19. Dixit, R. D. 1984. A Census of the Indian pteridophytes , Botanical survey of India, Calcatta.
20. Dixit R D & Sinha, B K 2001. Pteridophytes of Andaman Nicobar Islands, Bishen Singh Mahendra Pal Singh, Dehradun.
21. Dixit, R. D. and Singh, Shweta (2004). Medicinal Pteridophytes- An Overview. In Medicinal Plants Utilization and Conservation, Trivedi, P.C. (ed.), Aavishkar Publication, Distributors Jaipur. 268-297.
22. Dymock, William. 1890. Pharmacographia indica-A History of the Principal Drugs of Vegetable Origin met within British India. (Kegan Paul, London).
23. Fosberg, F R 1942. Uses of Hawaiian Ferns Amer. Fern J. 32 : 15-23.
24. Henry A N, Hosagoudor, V B & Kumar, R 1996. Ethnomedicobotany of the Southern and Western Ghats of India in S.K Jain (Ed) Ethnobiology in Human Welfare. 173-86. New Delhi.
25. Holdsworth, D K & Giheno, J 1975. A preliminary survey of highland medicinal plants. Sci.in New Guinea 3 : 198.
26. Jain, S.K. 1991. Dictionary of Indian folk medicine and ethnobotany. Deep Publication. New Delhi.
27. Jain, S. K. 2000. Human aspects of plants diversity. Economic botany 54 (4):459-470.
28. Jamir, N.S. 1997. Ethnobotany of Naga tride in Nagaland . Medicinal Herbs. Ethnobotany 9: 101-104.
29. Jacobs Marion Lee 1958. Index of plants of North Carolina with reputed medicinal uses. (Chapel Hill. North Carolina).
30. Jha, R R. & Verma S. K 1996 Ethnobotany of Sauria Paharia of Santhal Paragna, Bihar-I Medicinal plants. Ethnobotany, 8. 31-35.
31. Joshi, Pramila. 1997. Ethnobotany of pteridophytes of hilly district of Uttar Pradesh, India. Indian Fern J. 14. 14-18.
32. Kapur, S .K & Sarin, Y. K. 1977. Useful Medicinal Ferns of Jammu Kashmir. Indian Durgs. 14. 7: 136-140.
33. Kaushik, P & Dhiman, A. .K 1995. Common medicinal pteridophytes. Indian Fern J. 12.139-145.
34. Kirtikar, K R and Basu, B D 1918. Indian MedicinalPlants. Lalit Mohan Bose. Leader Road, Allahabad.

35. Kirn. & Kapahi, B. K. 2001. Ethnobotanical notes of some ferns and ferns allies of Jammu Kashmir, State. India. Indian Fern J. 18. 35-38.

36. Lloyd, Robert. M.1964.Ethnobotanical uses of California pteridophytes by Western Indian Amer. Fern. J. 54. 56-140.

37. May, L.W.1978. The economic uses and associated folklore of ferns and ferns allies. Botanical Review. 4. 4: 491-528.

38. Manickam, V. S. 1999. Medicinal ferns of India. Amruth (Feb): 3-9.

39. Nand Lal, S.P Singh, & Roy, S.P. 1982. Some medicinal ferns from South Andaman Island.Bull. Medi. Eth. Bota. Res. 3. (2-4): 178-185.

40. Nayar, B.K 1957. Medicinal Ferns of India. Bull Nat. Bot. Garden. 29. 1-36.

41. Niranjan, A.K.S & Dubey, J.P.1978. Some Medicinal Ferns from Pithoragarh Hills. J. Sci. Res.(B.H.U.).29. (10):183-189.

42. Pande, P. C & Bir, S. S.1994. Present assessment of rare and threatened vascular cryptogams (Pteridophytes) of Kumaun Himalayas and their conservation strategies. Indian Fern J. 11. 31-48.

43. Pande, P.C., P. Joshi & Pande, H.C. 1994. Studies on type Ethnobotany-II On less known Edible and economic ferns of Kumaun regions of W. Himal. Curr. Res. Pla. Sci. (Eds) Sharma T. A. Saini, S.S. Trivari, M. L. & Sharma, M. (Bishan Singh Mehandra Pal Singh. Dehradun).

44. Pande, P.C & Pangtey, Y. P. S. 1987. 1994. Studies on the Ethnobotany-I On less known Edible and economic ferns of Kumaun regions of W. Himal. J. Eco.Tax. Bot.2. (1):81-85.

45. Puri, G. S & Arora, R.K 1961. Some medicinal ferns from Western India. Indian For. 87. 179-188.

46. Puri, H. S 1970. Indian pteridophytes used in folk remedies. Amer. Fern. J. 56: 79-81.

47. Quisumbing, Eduardo.1951. Medicinal Plants of Philippines. Manila Dept. Arg. Nat. Res. Bull. 16 (2):174-176.

48. Roy, P & Gupta, H. N.1965.Cataka Samhita (A scientific synopsis) Nat. Inst. Sci.India.1-120 (History of Science in India publication).

49. Remero, John Bruno. 1954. The botanical Lore of California Indians. Vantage Press. N.Y.

50. Scully, Virginia.1970. A treasury of American Indian Herbs. Crown. N.Y.

51. Shankar Rama & Khare, P.K. 1992 Ethnobotanical observation on some ferns of Pachmarhi hills. Econ. Tax. Bot. Add. Ser.10: 97-100.

52. Shah, N.C and S.C. Singh 1990. Hitherto Unreported Phototherapeutic uses from Tribals of M.P. (India). Ethnobotany, 2: 91-95.
53. Singh, J. B 1969. Some medicinal ferns of Sikkim Himalayas. Indian J. Medi Res. 3 : 71-73.
54. Singh, J. B 1969-1970.Some medicinal ferns of Pachmarhi hills (M.P.) J. Sci. Res. BHU. 20 : 227-230.
55. Singh, S. 1993. Some edible ferns and fern-allies of North West Himalayas having promising food value Indian. J. Fer.16.(2):174-176.
56. Singh, K .K. & Maheshwari, J.K.1992. Folk medicinal uses of some plants among the Tharus of Gorakhpur district U.P.,India . Ethnobotany 4. 39-43.
57. Singh, K.K & Maheshwari, J.K.1989. Traditional herbal remedies among the Tharus of Baharaich district U.P., India Ethnobotany 1. 51-56.
58. Singh, K.K , Sahu, S.& Maheshwari, J.K.1989 Ethnobotanical uses of some ferns among the Tribals of Uttar Pradesh. India Indian. fern. J. 6 :63-67.
59. Singh, L S., Singh, P. K. & Singh E. Jadu. 2001. Ethnobotanical uses of some Pteridophytic species in Manipur, India Indian. fern. J. 18 : 14-17.
60. Singh, Shweta, Dixit, R. D. and Sahu, T. R. 2003. Some medicinally important Pteridophytes of Central India. *Int. J. For. Usuf. Magt.* 4(2): 41-50.
61. Singh Shweta, Dixit, R. D. and Sahu, T.R. (2005). Ethnobotanical use of Pteridophytes of Amarkantak (M.P.)". Indian Jour. Traditional Knowledge (CSIR) 4(4): Oct. 392-395.
62. Singh Shweta, Dixit, R. D. and Sahu, T.R. (2004). Ethnobotanical and indigenous knowledge of *Dryopteris cochleata* (Buch. Ham. ex D. Don) C. Chr. and *Tectaria coadunata* (Wall. ex Hook. et Grev.) C. Chr. among the tribal communities of Amarkantak. M.P." Jour. Bot. Soc. Univ. Sagar. 39: 113-117.
63. Smith, H.H.1924.Ethnobotany of Menomini Indians. Bull. Pub. Muse. City Milwaukee. (Green wood Press, Westport).4;1-174.
64. Srivastava, Kamini 2007. Important ferns in human medicine. Ethnobotanical Leaflets 11: 231-234.
65. Tiwari, S. D. N. 1964. Ferns of Madhya Pradesh. J. Indian Bot. Soc. 43. 431-452.
66. Turn, N & Marcus, Bell. A. M. 1971. The Ethnobotany of the Coast Salish Indians of Vancouver Islands. Economic Botany 25: 63-99.
67. Uddin, Md. G Mirza, M. M & Pasha, M. K. 1998. Bangaladesh J. Pl. Tax. 5 (2):29-41.

68. User, G. 1971.A dictionary of plants used by Man. (Hafner Press. N. Y.)

69. Upnof, J. C. Rh. 1968. Dictionary of Economic Plants. (Stechert Hafner. N.Y).

70. Vasudeva, S.M. 1999a. Ethnobotany of Pteridophytic flora of Pachmarhi, Tamia & Patalkot (Satpura Hills) Central India National Symposium on 50 years of Pteridodology in India in Retrospect and Prospect (Nov. 12-14) Jiwaji Uni. Gwalior. M.P. 30-31

71. Vasudeva, S.M. 1999b. Economic uses of Pteridophytes by the tribals of Tamia hills & Patalkot Valley Distt. Chhindwara of Madhya Pradesh. Bionotes 1(4): 81.

72. Vasudeva, S.M. 1999c. Economic Importance of Pteridophytes. Indian Fern J., 16:130-152.

73. Verma, P., A.A. Khan and K.K.Singh (1995). Traditional Phytotherapy among the Baiga Tribe of Shadol District of Madhya Pradesh. India. Ethnobotany.

74. Vyas, M.S & Sharma, B.B. 1988.Ethnobotanicalimportant of ferns of Rajasthan, Indigenous medicinal plants including microbes and fungi (Ed) Kaushik P. (Today's and Tomorrow Printers and Publications, New Delhi.).

75. Watt J. M & Beryer-Brandwijk, M.G.1962. The medicinal and poisonous plants of Southern and Eastern Africa. (2nd ed. Edinburgh, E & S. Livingstone).

Floral Diversity and their Conservation (2013), Editors: D.R. Khanna et al.
Published by Biotech Books.
ISBN: 978-81-7622-286-0 Pages: 291-298

29

Biodiversity of Prominent Cropland and Wasteland Medicinal Weeds in Amravati Region (India)

SHUBHANGI N. INGOLE
Department of Botany, Bar. R.D.I.K. and N.K.D. College, Badnera-Amravati

Amravati district is a part of Vidarbha in Maharashtra State (India) located in its north region. The present study deals with account of prominent medicinal weeds found in croplands, wastelands in Amravati region. It is not true that weeds are always harmful, they are in some way or the other beneficial to man. Even the modern man is also dependent on weeds for medicinal, ornamental and ceremonial purposes. Many wild weeds are still highly popular as medicine in India. The curative properties of various herbal medicines were referred in age-old suktas of Rigveda and Atharva Veda. Now world's attention turned to herbal medicines. Weeds are treasures of nature because of their influence on human life, hence they should not eradicated but should be put under control and conserved. In present work, total numbers of 57 useful weeds belonging to different angiospermic families out of which 35 have been recorded as medicinal. Besides current botanical names, common and local names, habit, uses and phenology have also been given for almost each taxon.

Key words: Weeds, Croplands, Herbal medicines, Amravati.

INTRODUCTION

Amravati district is a part of Vidarbha in Maharashtra State of India. It is located in its north region. Its area is about 12,212 sq. km; lying between 20032' to 21046' N and 76037' to 78027' E. The soil is loamy, black cotton soil for the largest area. PH values ranging from 6.5 to 8.5. Rainfall is moderate; climate is mostly hot. It is an innocent belief that all weeds are highly medicinal. In ancient India, treatises were written by eminent scholars on Ayurveda for the treatment of human beings and animals by using wild plants. These valuable works were destroyed during repeated foreign invasions and useful herbal knowledge fell into the hands of quacks and became unpopular. Even in modern times, about 200 medicinal herbs are used in Ayurveda. Native herbalists in villages are using many of the weeds folklore medicine since time immemorial. They are in some way or other beneficial to man. Even the modern man is also dependent on weeds for medicinal, ornamental and ceremonial purposes. Many wild weeds are still highly popular as medicine in India. The curative properties of various herbal medicines were referred in age-old suktas of Rigveda and Atharva Veda. Now world's attention turned to herbal medicines for their safety.

Weeds were not found at the start of terrestrial life on earth. As time passed, myriad of plant species evolved and clothes the naked rocky earth with green vegetation. In fact, mother nature has no concept of undesirable plant. For nature, all creatures are alike. The concept of weed and crop was developed by man depending upon his economic activity.

Knowledge of present status of its weed flora is essential as there are varied ecological and topographical sites, which sustain valuable biodiversity. Wild medicinal weeds exploration of this area has been attempted in the present paper.

MATERIALS AND METHODS

The plant collection trips were undertaken in such a way that all the seasons of the year could be covered. The plants were preserved by usual standard methods for further identification and confirmation.

A preliminary survey was made in Amravati region for wild medicinal weeds, in croplands and wastelands. Queries were made to local people regarding local names. Further survey was also made in industrially growing areas, dam sites, roadsides, etc. to determine the changing pattern of weed flora due to developmental activities and factors responsible for depletion of certain taxa. Standard floras, e.g., Flora of Maharashtra, literature and herbarium were referred for correct identification and nomenclature.

CONCLUSION

Witt (1916) published "Descriptive list of trees, shrubs, climbers and economic herbs of the Northern and Berar forest circles, Central provinces including Berar circle. Amravati region has received little attention. Most of the previous floristic studies of Amravati district were concentrated on Melghat (Witt, 1916; Patel, 1968; Dhore and Joshi, 1988; Bhogaonkar and Devarkar, 1999; etc.) Mirashi and Salpekar (1975) listed twenty-three species of ornamental plants from Amravati. Dhore (2002) carried out floristic survey of Amravati district with special reference to distribution of tree species.

In the present study, the total number of 57 prevalent wild medicinal weeds belonging to different angiospermic families have been recorded. In enumeration taxa are arranged alphabetically. Besides botanical names, common and local name, habit, uses and phenology have also been given for almost each taxon.

It has been found that many wild weeds play role of medicinal and therapeutic value. Hence, weeds are treasures of nature. Weed scientists should take the responsibility of awakening the common man to the economic and medicinal importance of wild weed vegetation. Instead of eradicating, wild weeds should be put under control and conserved.

Table 1: Cropland Wild Medicinal Weeds

S. No.	*Botanical Name*	*Common Name*	*Family*	*Uses*	*Habit*	*Flowers and Fruits*
1	*Chenopodium ambrosioides*	Doctor's chenopodium	Chenopodiaceae	Highly medicinal	Annual Herb	July-Oct.
2	*Chenopodium album* L.	Pigweed	Chenopodiaceae	Highly nutritious vegetable	Annual Herb	July-Oct.
3	*Convolvulus arvensis* L.	Morning glory	Convolvulaceae	Purgative and cattle food	Herb	Throughout year
4	*Chichorium intybus* L.	Blue daisy	Asteraceae	Vegetable, skin nourisher tonic herb	Herb	Throughout year
5	*Crotolaria verrucosa* L.	Dog pea	Fabaceae	Vermifuge and for skin diseases	Herb	Throughout year
6	*Cynodon dactylon* (L.) Pers.	Bermuda grass	Poaceae	Diuretic, fodder, Cooling effect	Perennial grass	Throughout year
7	*Eclipta alba* Hassk.	Bhrungraj	Asteraceae	Highly praised in Ayurvedic medicine as tonic, arrest premature graying of hairs	Herb	Throughout year
8	*Euphorbia hirta* L.	Physician's spurge	Euphorbiaceae	Highly medicinal, cures piles and stomach disorder	Herb	Throughout year
9	*Ipomoea triloba* L.	Dry land morning glory	Convolvulaceae	Cattle fodder rich in vitamin A, cures eye disorders	Crawling vine	Throughout year
10	*Lathyrus sativa* L.	Grass pea	Fabaceae	Cattle feed, seeds provide protein, used in manufacture of adhesives	Much branched with winged stem	Aug.-Oct.
11	*Lathyrus aphoca* L.	Wild pea	Fabaceae	Fodder, adhesives can be prepared from seeds, flowers are resolvent	Winter season annual with tendrils	Oct.-Jan.
12	*Melilotus indica* All	Sour clover	Fabaceae	Used as emollient nutritive fodder	Spreading Herb	Throughout year
13	*Phaseolus trilobus* Air.	Cat green gram	Fabaceae	Highly nutritious eaten by poor, useful in piles, tonic, leaves sedative	Prostrate stem wiry	July-Oct.
14	*Tridax procumbens* L.	Mexican daisy	Asteraceae	It has insecticidal and piscicidal properties, leaves used in dysentery, to arrest hemorrhage of wounds	Procumbent herb	Throughout year
15	*Vicia sativa* L.	Common vetch	Fabaceae	Used as fodder, rich in vitamins, proteins and minerals	Winter annual herb	Sept.-Dec.

Table 2: Wasteland Medicinal Weeds

S. No.	*Botanical Name*	*Common Name*	*Family*	*Uses*	*Habit*	*Flowers and Fruits*
1	*Abutilon vidicum* G. Don.	Wild lady's finger	Malvaceae	Stem yields good fibre, used for making ropes in rural areas. Bark contains astringent used for bowel complaints; leaves arrest breeding of piles. Roots found to be curing leprosy.	Much branched shrub	Through the year
2	*Acalypha indica* Linn.	Indian acalypha	Euphorbiaceae	Plant contains acalyphine which is good remedy for rheumatism. The root is good expectorant. Leaves past useful on skin disorders, excellent remedy for scabies	Erect herbaceous annual	July - October
3	*Achyranthes aspara* L.	Prickly-Chaff flower	Amaranthaceae	Having high medicinal value. Leaves used for worshipping Lord Ganesh, good remedy for piles, decoction oflus. Used in kidney troubles. Stem is used in cleaning teeth in rural areas. Shoot ash useful for toothache.	Erect, much branched perennial	July - October
4	*Acacia arabica* Willd.	Indian gum arabica	Mimosaseae	Leaves are astringent heal bone fractions and eye diseases. Flowers considered to be tonic. Gum is used in medicinal preparation of diabetes. Plant part aphrodisiac.	Small bushy tree	Through the year
5	*Agave Americana* L.	Century plant	Agavaceae	Slices of leaves are used as poultice. Leaf juice laxative, diuretic, emmnagogue and diaphoretic. Roots anti-syphilitic remedy for cancers.	Leaves very stout, in rosette .	Once in a Lifetime

contd...

S. No.	Botanical Name	Common Name	Family	Uses	Habit	Flowers and Fruits
6	*Ageratum conyzoides* L.	Beggar's ticks	Asteraceae	Decoction of plant is used in dysentery and gastrointestinal ailments	Erect annual herb, much branched densely covered with hairs'	July – November
7	*Amaranthus spinosus* L.	Thorny amaranth	Amaranthaceae	Boiled decoction useful in kidney troubles, antidote for snake bite. Ash of plant useful dyeing industry	Spiny, erect annual herb	July – October
8	*Argemone mexicana* Linn.	Prickly poppy	Papaveraceae	Seed oil useful in paint industry. Oil externally useful on skin diseases. The yellow juice useful in dropsy.	Erect annual sparingly branched annual	August – November
9	*Asteracantha longifolia* Nees.	Water spiny ball	Acanthaceae	Bitter root good remedy on rheumatism and urinogenital diseases. Seed, roots leaves used in jaundice, rheumatism, dropsy etc. Seed oil useful in paints.	Erect branched herb	Throughout The Year
10	*Cannabis sativa* L.	Indian hemp	Cannabinaceae	Preparations such as 'Bhang', 'charas' as intoxicants in India. Narcotic properties appreciated by pharmacopoeias, useful to induce deep sleep, in small dose goof appetizer. Powders of leaves useful on wounds, treating tumours.	Robust, erect annual herb	August – November
11	*Calandula officinalis* L.	Calendula	Asteraceae	Tincture extract of flowers is an excellent medicine to cure sores, ulcers, wounds and chronic vomiting. Fresh juice useful on jaundice	Erect annual pubescent	September – January
12	*Cleome chelidonii* Linn. F.	Wild mustard	Capparidaceae	Seeds used for culinary purposes in tribal areas. Oil useful in rheumatism	Less branched with small erect hairs	July – October

contd...

S. No.	Botanical Name	Common Name	Family	Uses	Habit	Flowers and Fruits
13	*Coccinia indica* WCA	Ivy gourd	Sucurbitaceae	Edible raw fruits useful as vegetable in rural areas. Leaves medicinals, useful in curing fevers.	Prostrate branched herb	July – November
14	*Croton sparsiflorus* Morong	Dog chilli	Euphorbiaceae	Young plant useful as green manure in rice nurseries. Leaf sap is used for healing cuts and wounds, is used to repel night insects, and prevent termite attack.	Erect, diffusely branched annual herb	August – December
15	*Indigofera trita* L. f.	Deep indigo	Fabaceae	Seeds are used as nutritive tonic, can be used as green manure.	Much branched erect undershrub	Throughout the Year
16	*Ipomoea maxima* (L.f.) G. Don.	Dry land	Ipomoea	Leaf juice is used as diuretic, good antidote for arsenic poison, tuber good tonic to uterus. Used as fence; stem as fuel wood in rural areas.	Erect, perennial with twining stem.	Throughout the Year
17	*Opuntia dillenii Haw.*	Serpent hood	Cactaceae	Fleshy fruit edible having cooling effect nutritious, sap is good hair tonic used as conditioners in herbal shampoos	Flattened green cactus with internodes	Occasional
18	*Portulaca oleracea* L.	Indian purslane	Portulacaceae	Leaves are sour with mucilage, good for curing malnutrition, leaves refrigerant	Low growing fleshy herb	Throughout the Year
19	*Prosopis juliflora DC*	Mesquite	Mimosaseae	Wood is of excellent burning quality. Wood used in house building, railway cross ties, etc. Root bark contains high tannin. Gum is used as emulsifying agents pods flour used in cakes and alcoholic beverages. Seeds protinaceous and excellent feed for cattles.	Shrub or small tree	Throughout the Year
20	*Tribulus terrestris* L.	Small caltrop	Zygophyllaceae	Used in several Ayurvedic medicines. Leaves and tender shoots are used as vegetable green leaves rich in calcium. Fruits are tonic, diuretic. Leaf juice is stomachic	Procumbent annual herb	July – December

REFERENCES

1. Bhogaonkar, P. Y. and V. B. Devarkar. 1999. Addition to the Flora of Melghat, Bulletin Number 7, Director, Project Tiger Melghat, Amravati Publication.
2. Dhore, M. A. 2002. Flora of Amravati District with Special Reference to Distribution of Tree Species, Amravati University Publication.
3. Dhore, M. A. and P. A. Joshi. 1988. Flora of Melghat Tiger Reserve, Technical Series No. 1, Directorate, Melghat Tiger Reserve Project Publication.
4. Mirashi, M. V. and J. B. Salpekar. 1975. Ornamental Plants at Amravati, The Botanique 6(4): 195-204.
5. Patel, R. I. 1968. Forest Flora of Melghat, Dehradun.
6. Singh *et al.*., 2000. Flora of Maharashtra State- Vol. 1, 2 and 3, Botanical Survey of India, Calcutta.
7. Witt, D. O. 1916. Descriptive list of Trees, Shrubs, Climbers and Economic Herbs of the Northern and Berar Forest Circles, Central Provinces, Allahabad.

Floral Diversity and their Conservation (2013), Editors: D.R. Khanna et al.
Published by Biotech Books.
ISBN: 978-81-7622-286-0 Pages: 299-319

30

Presence of Aerobial Mycoflora in and Around the Chennai City

R. Sampath Kumar[1], S. Dawood Sharief[2*] and M. Sekhar[3]
[1] M.Phil Scholar, School of Environmental Sciences,
P.G & Research Dept. of Zoology, The New College, Chennai
[2] Head, School of Environmental Sciences, P.G &
Research Dept. of Zoology, The New College, Chennai, India
[3] Technical Manager, Primer Scans and Labs, Chennai, India

Every day, the average person inhales about 20,000 liters of air. Every time we breathe, we risk inhaling dangerous chemicals that have found their way into the air. The effect of air pollution are diverse and numerous. Air pollution can have serious consequences on the health of human beings, and also severely affects natural ecosystems. The World Health Organization estimates that about two million people die prematurely every year as a result of air pollution, while many more suffer from breathing ailments, heart disease, lung infections and even cancer.

The varying climatical, geographical and social factors prevalent in India are ideally suited for the growth of many pathogenic fungi. The management of this fungal disease is very difficult. India is becoming the major 'stock-home' of AIDS patients in the South Asia, as a result, the fungal disease has reached noticeable proportion.

An investigation has been carried out to isolate and identify the fungi presence in the atmospheric air of Chennai City in 8 selected locations viz by settle plate method. The results revealed that 38 different species belonging to 22 genera of fungi were isolated from all the selected locations. The species such as *Aspergillus, Penicillium, Alternaria, Cladosporium* and *Candida* were commonly found in many samples. The number of colony forming units in a sample is also significant. The resulted fungal colonies and their health effects to human are discussed in the light of literature, the precautionary measures for such health problem are suggested.

Key words: Pollution, Settle Plate, *Aspergillus, Penicillium, Alternaria, Cladosporium, Candida*

INTRODUCTION

The occurrence of microorganisms in the atmosphere was speculated by Leeuwenhoek and in his letters to the Royal Society in 1680 (Dobell, 1932). Micheli (1729) first recorded the release of fungal spores in air but there were only a few detailed studies until the nineteenth century (Ainsworth, 1976). Pasture (1861) demonstrated visually the existence of air spora and pointed out that airspora must be measured while in suspension and not after deposition on surfaces. Aerobiology may be defined as the study of airborne particles of biological origin like Pollens, plant fragments, protozoa and miters and insects, their fragments. Fungal spores are thus only part of the total bioaerosol but nevertheless a very important part.

Airborne microorganisms associated with different occupations and work environments have been implicated in causing occupational respiratory diseases. Airborne particles which are biological in origin are commonly referred to as 'bioaerosols'. These 'particles' include bacteria, fungal spores, actinomycete spores, fern spores, pollen ,algae and antibiotics from biotechnology processes, endotoxins and excreta, proteins, enzymes and mycotoxins. These aerobes descend from different sources and their composition depends on the source material. Depending on their composition, the immunological reactivity of the exposed workforce and the circumstances of exposure, they may cause a range of occupational diseases through infection and also non-infectious and often immunological mechanisms. The effects of certain pathogenic bacteria, fungi and viruses have been recorded in some environments, but the more common effects of exposure to occupational bioaerosols are to mucous membrane irritation, bronchitis, allergic rhinitis, asthma, extrinsic alveolitis and organic dust toxic syndrome. Bioaerosols can be produced in many occupational environments such as farm, animal house, malting, storage and handling facilities for organic material, food processing plants and waste treatments plants and many others. Fungi produce large amount of spores which easily become airborne, thus constituting an important component in microbial aerosols.

A fungus is a member of a large group of eukaryotic organisms that includes microorganisms such as yeasts and molds, as well as the more familiar mushrooms. The discipline of biology devoted to the study of fungi is known as mycology, which is often regarded as a branch of botany, even though genetic studies have shown that fungi are more closely related to animals than to plants.

Around 1,00,000 species of fungi have been formally described by taxonomists, (Kirk *et al.*, 2008) but the global biodiversity of the fungus kingdom is not fully understood (Mueller, Schmit, 2006). On the basis of observations of the ratio of the number of fungal species to the number of plant species in selected environments, the fungal kingdom has been estimated to contain about 1.5 million species; (Hawksworth, 2006) and a recent (2011) estimate suggests there may be over 5 million species (Blackwell (2011). In mycology, species have historically been distinguished by a variety of methods and concepts.

Some fungi can cause serious diseases in humans, several of which may be fatal if untreated. These include Aspergilloses, Candidoses, Coccidioidomycosis, Cryptococcosis, Histoplasmosis, Mycetomas, and Paracoccidioidomycosis. Furthermore, persons with Immuno-deficiencies are susceptible to disease by genera such as *Aspergillus*, *Candida*, *Cryptoccocus* (Hube, 2004; Nielsen, Heitman, 2007; Brakhage, 2005), *Histoplasma* (Kauffman, 2007) and *Pneumocystis* (Cushion, 2007). Other fungi can attack eyes, nails, hair, and especially skin, the so-called dermatophytic and keratinophilic fungi, and cause local infections such as ringworm and athlete's foot (Cook, 2008). Fungal spores are also a cause of allergies, and fungi from different taxonomic groups can evoke allergic reactions (Simon-Nobbe, 2008).

Considering the above aspects one such place in south India, Chennai (Second most populated city in India) has all above environments (within) it. This is the most vulnerable place for aerobes, because it is well known that many places of the City is being maintained un-hygienically, which gives room for the collection of pathogenic aerobes. It is also the venue where tourist and foreign peoples visit often for historic places. Air borne microbes present there may cause diseases in them and also affect us. Hence the present study was conducted with the following aim and objectives.

- To analyse for the presence of atmospheric molds in Chennai, with particular reference to fungi.
- Health effects associated with presences of molds in Chennai City.
- To suggest precautionary measures associated with present of molds.

MATERIALS AND METHODS

Collection of samples

Samples were collected by exposing petriplates containing nutrient agar medium to atmospheric air up to 30 minutes at different selected locations of the Chennai city. The plates were placed at a height of 1 meter above ground level. The climatical conditions of the places influenced the presence of aerial moulds.

Sample collection: I

The samples were collected from 11:00 to 11:30 am at all selected places of Chennai city - Ambattur, Anna Nagar, Guiney, Mylapore, Porur, Royapuram, Sanatorium and Thuraipakkam during the month of February, March and April 2011 every fortnight. At the time of the sample collection the climatical condition was quite normal, the temperature was 29.0^{0}C, humidity was 46%, and wind was flowing from East and North-East direction at the speed of 4 km per hour (Table: 1. courtesy data from Metrological Department of Chennai).

Table 1: Isolated fungal colonies from sites places of Chennai during the month of Feb2011 to Apr-2011. ,

Species	P-I	P-II	P-III	P-IV	P-V	P-VI	P-VII	P-VIII	Total
Absidia corymbifera	-	3	-	-	-	-	-	2	5
Alternaria alternate	3	5	5	12	8	4	2	8	47
Alternaria dinothica	-	4	-	-	-	2	-	-	6
Aspergillus chealieri	-	2	1	-	-	-	2	-	5
Aspergillus flavipes	10	15	-	2	6	-	-	-	33
Aspergillus flavus	7	13	10	13	12	15	4	8	82
Aspergillus fumigatus	-	5	-	6	-	7	2	12	32
Aspergillus glacus	-	-	8	2	15	2	-	-	27
Aspergillus japanicus	3	2	4	-	-	-	10	10	29
Aspergillus niger	10	21	15	10	61	17	15	38	187
Aspergillus nidulace	-	-	20	6	17	18	6	21	88
Aspergillus tamarii	3	-	-	-	3	-	-	-	6
Aspergillus terreus	-	-	3	3	-	-	5	-	11
Aspergillus ustus	-	4	-	-	2	-	-	-	6
Aspergillus versicolor	4	3	7	-	-	4	-	2	20
Blastosporium canis	-	-	1	-	-	4	-	-	5
Canadidia albicans	20	8	10	14	11	7	-	17	87
Cladosporium	9	3	6	-	-	-	10	11	39
Curvularia clavata	-	2	-	1	-	3	-	2	8
Curvularia lunata	-	-	7	-	5	6	3	6	27
Eercilla nidulan	1	-	-	-	-	-	2	-	3
Epidermophyton	-	-	-	2	-	3	-	-	5
Fusarium	9	6	18	8	9	-	17	7	74
Histoplasma	-	-	-	4	-	-	6	-	10
Humicola grisea	-	3	-	2	-	-	-	-	5
Microsporium	-	-	-	4	-	5	-	2	11
Mucor	7	3	5	3	8	-	-	5	31
Nigrossporium spaerica	-	-	-	2	3	4	2	4	15
Penicillium citrinum	13	2	7	7	10	7	7	10	63
Penicillium funiculosum	14	5	7	3	2	-	7	-	38
Penicillium marneffei	10	6	6	-	-	3	-	3	28
Rhizopus	2	11	5	-	17	-	2	2	39
Scolecobasicum	-	-	-	-	25	3	-	-	28
Scopularia brevicauli	-	-	-	-	-	-	2	-	2
Trichoderma	4	8	17	6	-	1	-	4	40
Trichophyton	-	-	3	-	-	5	-	11	19
Trichotheium roseum	4	-	-	1	-	-	2		14
Total	**133**	**134**	**165**	**111**	**214**	**120**	**106**	**192**	**1175**

Sampling was carried out at the following sites of Chennai.

Site: 1 – **Ambattur,** Site: 2 – **Anna Nagar,** Site: 3 – **Guindy** Site: 4 – **Mylapore,** Site: 5 – **Porur,** Site: 6 – **Royapuram,** Site: 7 – **Sanatorium,** Site: 8 – **Thuraipakkam.**

As mentioned above samples were collected from all the sites.

Sample collection II: Feb-25, 2011 at 11:00 to 11:30 am (Temperature – 30.2^{0}C; Humidity – 52%).

Sample collection III: Mar-10, 2011 at 11:00 to 11:30 am (Temperature – 32.8^{0}C; Humidity – 31%).

Sample collection VI: Mar-25, 2011 at 11:00 to 11:30 am (Temperature – 31.8^{0}C; Humidity – 58%).

Sample collection V: Apr-10, 2011 at 11:00 to 11:30 am (Temperature – 32.2^{0}C; Humidity – 60%).

Sample collection VI: Apr-25, 2011 at 11:00 to 11:30 am (Temperature – 31.2^{0}C; Humidity – 67%).

ISOLATION OF AERIAL MOLDS BY SETTLE PLATE METHOD

The number of microorganisms deposited onto the agar surface of the plate over the period of exposure was determined by incubation of the agar plates at 25°C for 5- 7 days and counting colonies that develop. The results are expressed as number of Colony Forming Units (CFUs) per unit time (Checklist of settling plates (QU-04-0003-FRM). The counted colonies are further characterized to genera or species. Higher numbers of CFUs and/or presence of potential pathogenic or toxigenic mold such *Aspergillus fumigatus* and *Stachybotyrs chartarum* are indicators of a problem.

A number of methods can be used to test air for mold or other microbial contamination. One of the oldest methods of testing air for microbial contamination is the settle plate's method. Settle plate method is semi-quantitative; it is still considered a useful method. In industries such as food, pharmaceutical and cosmetics the method is used to assess the likely number of microorganisms depositing onto the product or surface in a given time. The method involves opening and exposing Petridishes containing agar medium suitable for growth of microorganisms. The agar plates are left open at table-top level at selected locations for half-hour to 4 hours. This allows mold spores and fragments to settle onto agar media by gravity. The recommended size of solid media is 90 mm in diameter (approximate internal area 64 cm^2) for settle plates.

1. Media

Sabouraud Dextrose Agar

Sabouraud Dextrose Agar is used for the cultivation of fungi, following the United States Pharmacopeial Convention (2007); Directorate for the Quality of Medicines of the Council of Europe (EDQM) (2007); Japanese Pharmacopoeia (2007); Sabouraud, (1892); Jarett, and Sonnenwirth (eds.), (1980); Georg, Ajello, and Papageorge, (1954); Curry, Graf, and McEwen, Jr. (eds.), (1993).

1.2 Composition of Sabour aud Dextrose Agar (SDA)

Enzymatic Digest of Casein	5 g
Enzymatic Digest of Animal Tissue	5 g
Dextrose	40 g
Agar	15 g

Final pH is adjusted to 5.6 ± 0.2 at 25°C

1.3 Procedure

65 g of the medium was suspended in 1000ml distilled water, mixed thoroughly, heated with frequent agitation and boiled for 1 minute for complete dissolution of the medium. The pH was adjusted to 5.6 and sterilized for 15 minutes in an autoclave at 121°C (15 lbs pressure) and used for the culture studies.

Limitations of the media

1. Some strains may be encountered that grow poorly or fail to grow on this medium.
2. Antimicrobial agents are to be added into a medium to inhibit bacteria and also certain pathogenic fungi.
3. Overheating a medium is to be avoided, which may result in a soft medium in acidic pH.

2. Procedure for Preparation of Settle plates

i. Approximately 25 ml of sterile SDA agar is poured into sterile Petridish (90mm in diameter), allowed for solidifying (Checklist of settling plates (QU-04-0003-FRM).

ii. The plates were inverted and incubated, for 24 hours at 30-35°C to verify their sterility.

iii. The bases of plates (the part containing the media) were labelled with numbers provided with the name, location, date, time.

3. Settle plates Sampling

a) The plates were transferred into the area, where the test to be conducted.

b) Plates were placed at table-stand height (1 meter from ground).

c) The lids of plates were raised to expose the surface of the medium, the lid was rested on the very edge of the plate so that the entire agar surface is completely exposed, and care was taken not to put fingers on plates (Checklist of settling plates (QU-04-0003-FRM); Rules and Guidance for pharmaceutical manufacturers and distributors, 2002).

4. Sample locations

Sampling was undertaken in the following locations specified.

Site. No	*Location*	*Media exposure*	*No of plates exposure*	*Exposure time*
1	Ambattur	SDA	6	30 minutes
2	Anna Nagar	SDA	6	30 minutes
3	Guindy	SDA	6	30 minutes
4	Mylapore	SDA	6	30 minutes
5	Porur	SDA	6	30 minutes
6	Royapuram	SDA	6	30 minutes
7	Sanatorium	SDA	6	30 minutes
8	Thuraipakkam`	SDA	6	30 minutes

5. Incubation

The exposed petriplates containing SDA were immediately closed and transported to the laboratory for incubation. The plates were incubated at room temperature (28 ± 2^{0}C). After incubating for the period of 5 days the colonies were counted and identified.

6. Fungal identification

Identification of fungi by observation method (James, Cappaccino, Natalie Sherman, 1999).

LACTO PHENOL COTTON BLUE

The stain used for identification of fungi is Lactophenol cotton blue. The identification of molds is based on the shape, method of production, and arrangement of spores. (McInnis,1980; Larone,1995, Murray,1999; Clark,1981; Suarez Peregrin,1972; Emmons, 1977; Baron,1990; Forbes, *et al*.,1998; August, *et al*.,1990; Isenberg, Koneman, *et al*., Philadelphia,1997; Larone, 1993; Kwon-Chung,1992).

PROCEDURE

A Drop of Lactophenol was placed on the centre of a clean microscope slide. Small portion of growth was removed from the midway between the colony centre and edge, with a long-handled inoculating needle. The material was placed in Lactophenol solution. Gently teas the fungus apart by dissecting needles for thin spreading in the Lactophenol. Coverslip is placed on the edge of the Lactophenol and slowly lower it with a sharp pointed object for avoiding trapping of any air bubbles under the coverslip; Excess Lactophenol was removed from the edges of the coverslip by using blotting paper. The preparation was examined under high power and low power magnification with the aid of dissecting microscope and observations were recorded.

RESULTS

Isolation and identification of fungal population present in the Atmospheric air of Chennai was achieved by randomized sampling at 8 selected sites from the month of February 2011 to April 2011. The reported fungal colonies were tabulated based on the difference in location and duration.

Site I:

A total of 133 colonies were recorded with 11 genera from Ambattur. Among this fungal population, genus *Aspergillus* (28.9%) was represented with 7 species, *Penicillium* (28.9%) 3 species and *Alternaria* (2.3%), *Candidia albicans* (15.4%), *Cladosporium & Fusarium* has same number of colonies (7.03%) while *Tricoderma* (3.12%) is represented by single species.

Site II:

A total of 131 colonies were recorded with 22 fungal species belongs to 12 genera. From this fungal population, *Aspergillus* contributed (50%) represented by 7 species, *Penicillium* (10%) with 3 species and *Alternaria* had 2 species recorded. The remaining species like *Candidia* (6.5%), Absidia corymbifera (2.4%) *Fusarium* (4.9%) was represented by single species.

Site III:

A total of 165 colonies were recorded with 21 species belongs to 12 genera. Genus *Aspergillus* contributed 41.8% of the total colonies, represented by 7 species, followed by *Penicillium* (12.5%) with 3 species. The remaining was reported by *Fusarium* (11.2%), *Rhizopus* (9%), *and Candidia albicans* (6.25%) with single species.

Site IV:

111 colonies were recorded with 21 species of fungal colonies belonging to 14 genera. From this genus *Aspergillus* has contributed 37.5% with 7 species, *Penicillium* 8.9% with 3 species. The remaining genera were represented with single species.

Site V:

A total of 214 colonies were recorded with 17 species belongs to 10 genera. *A.niger* was reported from all the samples of site-V, followed by *A.niger*, *Tricoderma* contributed 12.2%, *A.nidulans* and *Rhizopus* contributed 8.3% over the other species of fungal colonies (table 6).

Site VI:

A total of 120 colonies were recorded with 20 species belongs to 12 genera. From this fungal population, genus *Aspergillus* contributed (57.2%) represented by 6 species. *Penicillium* (9%) with 2 species, *Curvularia* (8.1%) and *Alternaria* represented with 2 species. The remaining species recorded only one species from each.

Site VII:

A total of 106 colonies were recorded from the atmospheric air of site-VII has represented by 12 species belongs to 14 genera of fungi. In this *Aspergillus* alone contributed 40.5% with 7 species; *Penicillium* contributed 8.9% with

3 species, Curvularia with 2 species and the remaining was recorded with single species.

Site VIII:

A total of 183 colonies were recorded with 21 species of fungal colonies belonging to 14 genera. The genus *Aspergillus* has contributed 48% over other colonies, followed by *Penicillium* and *Aspergillus*. The remaining genera were represented with single species.

DISCUSSION

Wide spectrums of fungal diseases such as superficial, subcutaneous and systemic mycoses are known to be prevalent in India. The varying climatic, geographic and social factors prevalent in India are ideally suited for the growth of many of pathogenic fungi. Because of the high rate of illiteracy, still major sections of the population believe in penance for cure of a diseases rather than treatment. Hence, the managements of many of the fungal diseases have become difficult. India is becoming the major 'stock-home' of AIDS patients in the entire south Asia, as a result, the fungal diseases has reached noticeable proportions. This sends a clear warning to the Indian medical fraternity that if early diagnoses of fungal diseases are not made, mycoses may surpass the diseases caused by all other pathogens put together in future.

Fungi produce a large number of spores which easily become airborne and can infect man and animals and can cause allergy and mycotoxicosis (Lacey, 1996). Providing protection to people from health hazards exposed to biological agents, especially microbes becomes an increasingly important factor in health aspect. There is a need to enumerate microorganisms at different place to access the exposure to microbial risk.

The interest in fungal exposure has increased over the last decades. It is largely because it is now appropriately recognised that exposures to biological agents in both the residential and occupational environment are associated with a wide range of adverse health effect with major public health impact, including contagious infectious diseases, acute toxic effect, allergic and cancer. Despite the recognition of the importance of fungal moulds exposed on human health, the precise role of biological agent in the development and aggravation of symptoms and diseases is only poorly understood. The present study is conducted with the objective of analyzing the fungal presence in the atmospheric air of Chennai City, which is highly congested populated and for the heaviest vehicular movements, causing various kinds of lung diseases, skin, eye, etc.

There is different genus of fungi isolated from 8 selected sites of Chennai. Among this genera *Aspergillus* contributes (46%) which is comparatively high over other genus, and it is due to its capability to grow in nutrient-depleted environments. Similar observation was recorded by Ghani in the year of 1977, from atmospheric air of vegetable market. The predominance of *Aspergillus* in the air of leather industry at Bangalore was observed by Manjunath and Babu, 1997.

The genus *Aspergillus* is represented by 12 different species, among these species *A.niger* alone contributed to 16% over the other genus. This observation is supported by the studies of Pugalmaran (1997). In his study 70.8% of *A.niger* was isolated from air of leather store house, followed by *A.niger, A.nidulance* and *A.flavus* contributed 7%, it may be noted that similar observation were recorded by Udaya Prakash (1997). Remaining species of *Aspergillus* were reported with very low number of colonies from the all the aerobial samples of Chennai.

The *Aspergillus* species are the main causative agent for variety of disease, particularly Aspergillosis. The most common forms are Allergic bronchopulmonary Aspergillosis (ABPA), Pulmonary Aspergilloma and invasive Aspergillosis. Aspergillosis develops mainly in individual who are immunocompromised. The most common causative agent of these diseases is *Aspergillus fumigatus*. These diseases can be treated with medicines like Voriconazole and Liposomal amphotericin B for 6- 9 months preferably as suggested by Herbrecht, 2002. Aspergilloma can be cured by treating with Itraconazole and the treatment for ABPA consists of Corticosteroids and antifungal medications (Longmore, 2004).

In this present study 3 species of *Penicillium* were isolated with 11.2% of contribution. This is the second highest contribution of this study. Among this species *P.citrinum* (5.4%) was recorded with high contribution followed by *P.funiculosum* (3.2%) and *P.marneffei* (2.4%). This is in concurrence with studies of Pugalmaran, 1997. Penicilliosis is an infection caused by *Penicillium marneffei* (Desakorn, 1999). It can be treated with antifungal agents like Ketoconazole, Itraconazole, Miconazole, Flucytosine and Amphotericin B according to Wu, 2008.

During this study period 2 species of *Alternaria* were recorded with 4% of colonial contribution over other species, specially from the Mylapore site 12 number of colonies were recorded. Followed by *Alternaria, Curvularia* species were reported with 2 species by the contribution of 3.6%. Wound infections, mycetoma, onychomycosis, keratitis, allergic sinusitis, cerebral abscess, cerebritis, pneumonia, endocarditis, dialysis-associated peritonitis, and disseminated infections developed due to *Curvularia* spp. flucytosine yielded very high MICs for *Curvularia.*

Candida albicans was reported very high number from almost all the sampling sites. *Candida* alone recorded with 7.4% of its contribution over other species. Candidiasis is fungal diseases caused by any of *Candida* species. In this present study *Candida* was reported with high number. Candidiasis is the superficial infection of skin and mucosal membranes. According to Craigmill (1991), this can be treated with antimycotics – Clotrimazole, nystatin, fluconazole and ketoconazole. Infection of the oesophagus is caused by *Candidia albicans*. This disease occurs in patients who are in immunocompromised states. It is also known as monilial oesophagitis. According to Hamza (2008), this is treated by a single dose of fluconazole (750mg).

Followed by *Candidia* species is *Fusarium*, which was recorded with the contribution of 6.36%. Similar observations were reported by Meshram in 1999, from a study on the incidence of *Fusarium* in the air of fruit and vegetable market. *Fusarium* spp. is causative agents of superficial and systemic infections in humans. *Fusarium* strains yield quite high MICs for flucytosine, ketoconazole, miconazole, fluconazole, and posaconazole.

It is interesting to note that *Mucor, Rhizopus, Trichoderma* and *Absidia corymbifera* species has been recorded almost in same number of total colonies from various places of study. *Nigrosporium spaerica* and *Trichophyton* has contributed with 1% of the total. Verma *et al.*. (1994) has recorded, *Nigrosporium* species followed by *Aspergillus* and *Cladosporium* at Jabalpur, which in concurrence with our present studies. The genus *Trichophyton* is the causative agent of athlete's foot or ring worm of the foot (Rapini, 2007). It is a fungal infection of the skin that cause scaling, flaking, and itch of affected areas. This can be controlled by applying Allyamines and Azoles (Crawford 2007).

Emercilla, Epidermophyton, Histoplasma, Humicola, Microsporium, Scolecobasicum, Scopularia and *Trichotheium* were recorded less than the

1% of the percentage contribution over the other species. More over this species are not reported consequently from all the sites. In this, reported species *Epidermophyton, Microsporium* and *Trichophyton* are deramtophytes which may cause skin, hair and nails infections. This infectious are called Tinea, Onycomycosis and Kerion development. Tinea (nail) requires oral treatment with Terbinafine, itraconizole. Tinea capitis (scalp) and it must be treated orally with Griseofulvin for 2-3 month. Pedis tinea is usually treated with tropical medicine.

Even though *Histoplasme* species is reported very low, it is one of the important causative agents of Histoplasmosis and it refers to Darling disease, Cave disease, Ohio valley, Reticuloendotheliosis, Spelunker's Lung and Caver's disease. Histoplasmosis is common among AIDS patients, because of their suppressed immune system (Cotran 2005). This can be cured by the treatment with Amphotericin B, followed by oral suspension of Itraconazole (Barron, 2008).

High humidity allows fungal growth (mainly of *Penicillium* and *Aspergillus*), with concomitant release of conidia and fragments into the atmosphere. The number of isolated fungal colony is in direct proposition with Humidity of the atmosphere, by comparing the number of colonies with report from Metrological Department of Chennai.

Present study reveals that high number of disease causing fungal colonies from the atmospheric air in different part of Chennai may be due to high humidity, which corresponds with the data obtained from Metrological Department Chennai. To prevent from this fungal disease, we should keep our environment free from piles of leaves and decaying vegetable matter, Avoid lawn mowing or raking and limiting the amount of time spend outdoor during the times of high mold spore concentration (high humidity); wearing a face mask will help very much to avoid fungal spores.

Further to avoid infections the rural/urban high ways where in most of the solid /liquid waste released from various resources must be cleaned periodically, with disinfectant, so that spores do not form and fly along with fly ash, which in future could cause health hazard to the public. The people of Chennai should take their own precautionary measures by covering their face by suitable masks so as to avoid entry of the fungal spores either through nasals or through mouth or infect the eyes.

Most important aspect of spreading of fungal spores is by "Living amids garbage, inhaling the stench and exposing to all kinds of poisonous substance make people vulnerable to all diseases, where in the fungus spread. Hence precautionary measures have to be ensured to cover with tarpaulin, instead of netted wire meshes while the garbage is transported from the City to dumping yards. Though the civic authorities are taking some measures, still the way in which the garbage is being produced do not match with the cleaning ability of the area more and more cleaning vehicle has to introduce to reduce the menace and people should also be taught about garbage separation by handbills and awareness campaign has to periodically conducted through Schools, College, NSS, NCC and NGO's.

It may be noted from this study that aerobes are present in water, soil and in air causing various kinds of diseases to the mankind. Since this study pertains to fungus in air, the causes and the remedial measures are highlighted. It can also be ascertained from this study that the sites, which are highly polluted by aerobes are:

PORUR > THURAIPAKKUM > GUNIDY > AMBATTUR > ANNANAGAR > ROAYPURAM > MYLAPORE > SANATORIUM

Because these areas, Porur on Poonamallee highways (NH-4) and Chennai Bypass is surrounded by number of chemical industries, manufacturing plants, residential colonies and vegetable markets and solid waste dumping sites, and high in number of fungal colonies, followed by Thuraipakkam, on the old Mahabalipuram road (OMR), where in road widening and buildings are coming up and the corporate companies are making lot of expansion for water ways, sewage properties and wading of the road, followed by Guindy amids Metro railway project and Ambattur being industrial hub, while Anna Nagar being highly populated, followed by Royapuram for Truck and Trailers, Mylapore for public movements and Sanatorium by National high way. Chennai is slowly becoming highly populated, congested with heavy vehicular movements, due to construction of high bridges and this indicates that atmosphere and atmospheric air is contain fungal spores causing diseases, which has to take care by the Government and as well as public.

Slide I: Fungal identified in the Atmospheric Air of Chennai City (Feb to April 2011)

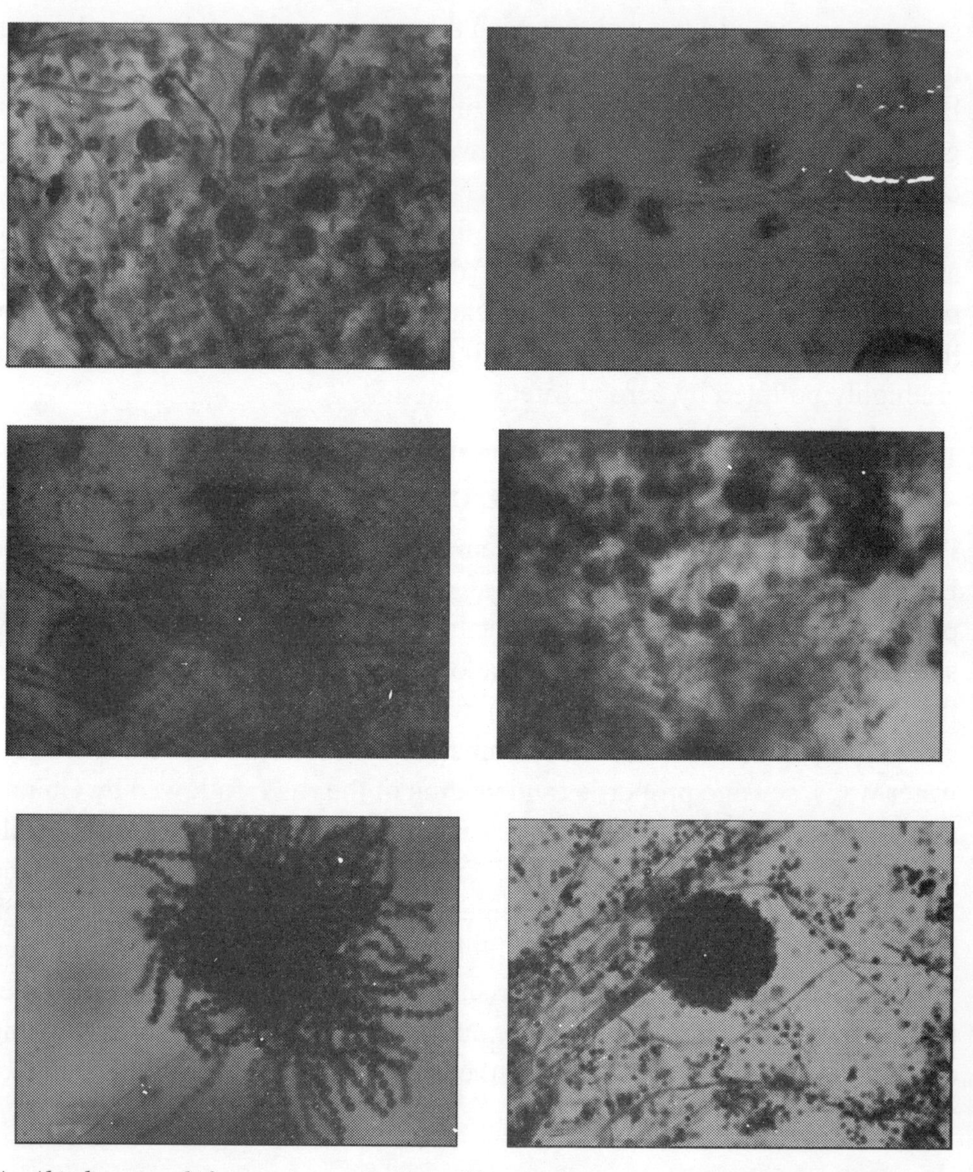

A. *Abisdia corymbifera* B. *Asperillus chealieri* C.*Aspergillus flavus*
D.*Aspergillus fumigatus* E. *Aspergillus glaucus* F. *Asperigllus japanicus.*

Slide II: Fungal identified in the Atmospheric Air of Chennai City (Feb to April 2011)

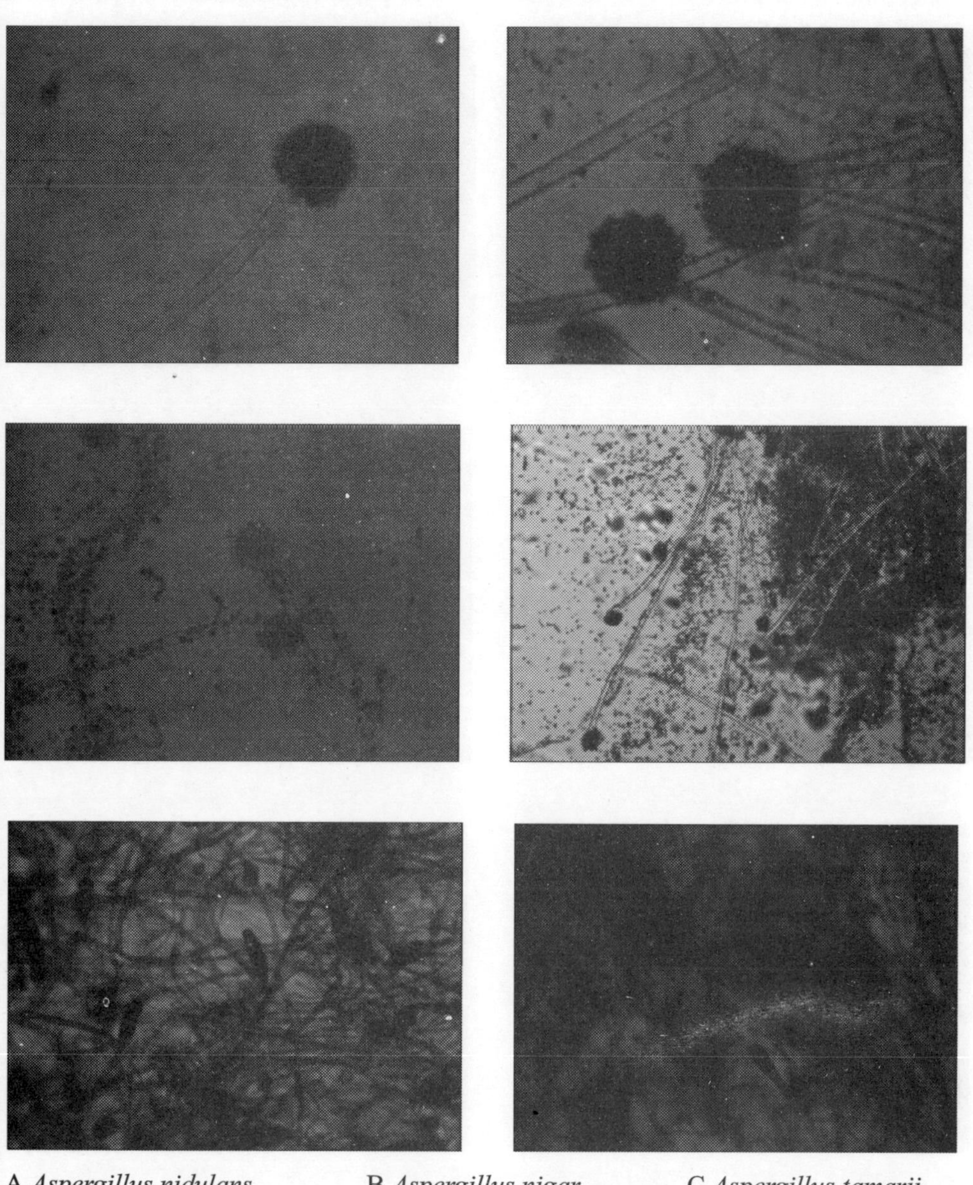

A.*Aspergillus nidulans* B.*Aspergillus nigar* C.*Aspergillus tamarii*
D.*Asperigllus versicolor* E.*Alternaria alternate* F. *Alternari dinapthicus.*

Slide III: Fungal identified in the Atmospheric Air of Chennai City (Feb to April 2011)

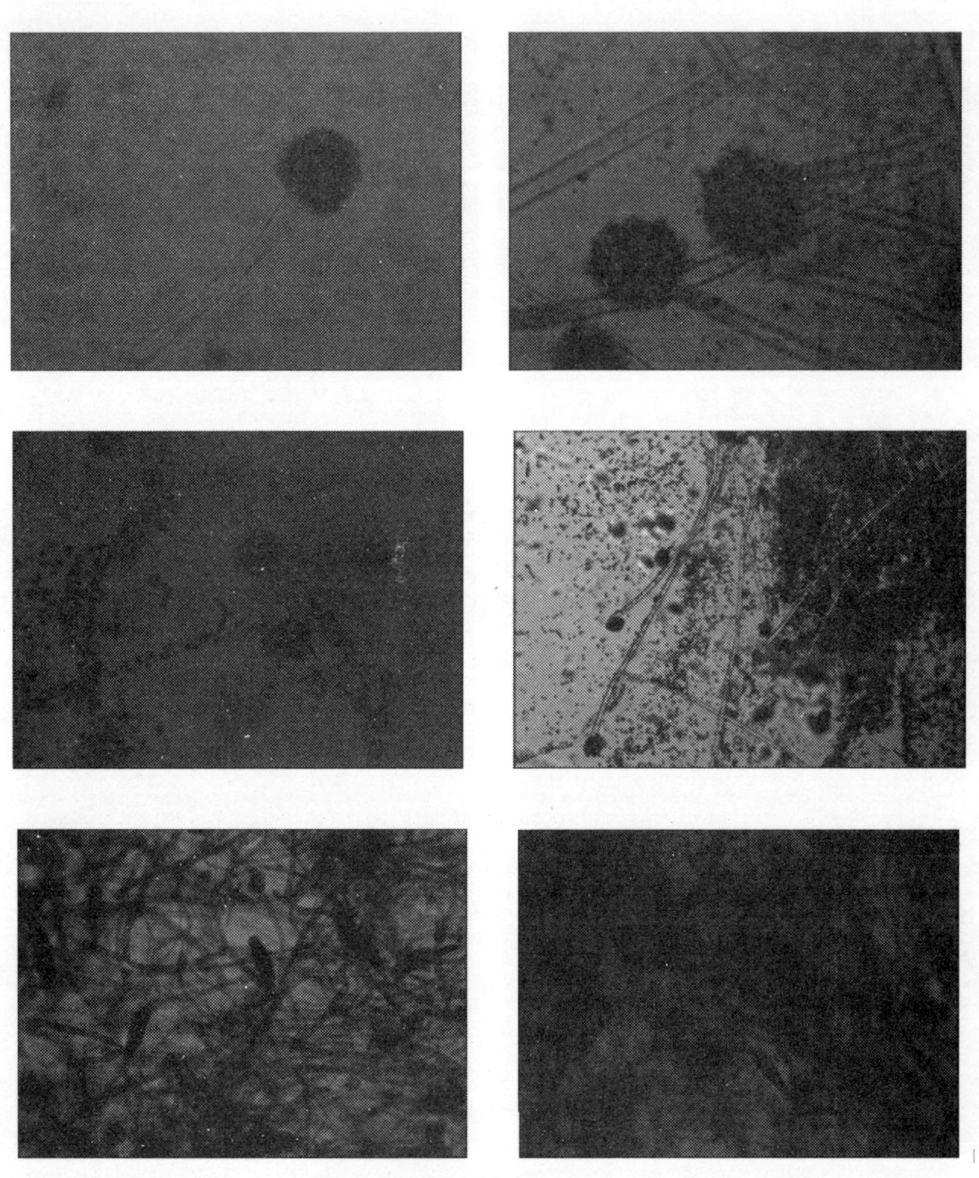

A.*Fusarium* sp. B. *Monilia sitophila* C.*Mucor* spp.
D. *Pinicillium citrnum* E. *Humicola grisea* F. *Nigrospora shaeica*
G.*Scolecobasidium* sp. H.*Trichothecium roseum*

REFERENCES

1. 1. Anisworth, G.C. 1976. Introduction to the history of Mycology. Cambridge University Press, Cambridge, pp. 359.
2. 2. August, M.J., *et al.*, Cumitech 3A; Quality Control and Quality Assurance Practices in Clinical Microbiology, Coordinating ed., A.S. Weissfeld. American Society for Microbiology, Washington D.C., 1990.
3. 3. Baron E.J., and Finegold S.M., Bailey & Scott's diagnostic microbiology. St. Louis: Mosby Company (1990).
4. Blackwell M. (2011). "The Fungi: 1, 2, 3 ... 5.1 million species?" (PDF). *American Journal of Botany* 98 (3): 426–438. doi:10.3732/ajb.1000298. PMID 21613136. http://www.amjbot.org/cgi/reprint/98/3/333.
5. Brakhage AA. (2005). "Systemic fungal infection caused by Aspergillus species: epidemiology, infection process and virulence determination". Current Drug Targets 6 (8): 875 - 886.
6. Clark G., "Staining Procedures", 4th Ed., Williams and Wilkins , Baltimore 362 (1981)
7. Cook GC, Zumla AI. (2008). *Manson's Tropical Diseases: Expert Consult.* Edinburgh, Scotland: Saunders Ltd. p. 347. ISBN 1-4160-4470-1.
8. Craigmill A. (1991). "Gentian Violet Policy Withdrawn". Cooperative Extension University of California – Environmental Toxicology Newsletter 11 (5).
9. Curry, A. S., J. G. Graf, and G. N. McEwen, Jr. (eds.). 1993. CTFA Microbiology Guidelines. The Cosmetic, Toiletry, and Fragrance Association, Washington, D.C.
10. Desakorn V, Smith MD, Walsh AL, *et al.*(1999). Journal of clinical microbiology 37 (1): 117-21.
11. Directorate for the Quality of Medicines of the Council of Europe (EDQM). 2007. The European Pharmacopoeia, Amended Chapters 2.6.12, 2.6.13, 5.1.4, Council of Europe, 67075 Strasbourg Cedex, France.
12. Emmons C.W., Binford C.H., Utz J.P., and Kwon-Chung K., J. Medical mycology. 4th ed. Philadelphia: Lea &Febiger (1977)
13. Forbes, B.A., *et al.*, Bailey and Scott's Diagnostic Microbiology, 10th ed. C.V. Mosby Company, St. Louis, MO, 1998.
14. Georg, L. K., L. Ajello, and C. Papageorge. 1954. Use of cycloheximide in the selective isolation of fungi pathogenic to man. J. Lab Clin. Med., 44:422-428.

15. Hamza OJM, Matee MIN, Bruggemann RJM, *et al.* (2008). "Single-dose fluconazole versus standared 2-week therapy for orophayngeal candidiasis in HIV-infected Patients". Clin Infect Dis 47 (10): 1270 – 1276.
16. Hawksworth DL. (2006). "The fungal dimension of biodiversity: magnitude, significance, and conservation". *Mycological Research* 95 (6): 641–655. doi:10.1016/S0953-7562(09)80810-1.
17. Herbrecht R, Denning D, Patterson T, Bennett J, Greene R, (2002). "Voriconazole versus amphotercin B for primary therapy of invasive aaspergillus". N Engl J Med347, (6). PMID 12167683.
18. Hube B. (2004). "From commensal to pathogen: stage- and tissue-specific gene expression of *Candida albicans*". *Current Opinion in Microbiology* 7 (4): 336–341. doi:10.1016/j.mib.2004.06.003. PMID 15288621.
19. Isenberg, H.D.,Clinical Microbiology Procedures Handbook, Vol. I & II. American Society for Microbiology, Washington D.C.
20. Japanese Pharmacopoeia. 2007. Society of Japanese Pharmacopoeia. Amended Chapters 35.1, 35.2, 7. The Minister of Health, Labor, and Welfare.
21. Jarett, L., and A. C. Sonnenwirth (eds.). 1980. Gradwohl's and parasitic infections, 7th ed. American Public Health Association, Washington, D.C.
22. Kwon-Chung, K.J. and J.E. Bennett, Medical Mycology. Lea and Febiger, Malvern, PA, 1992.
23. Lacey, 1996. Spore dispersal-its role in ecology and disease. The British Contribution to Aerbiology. Myco. Res. 100 (6): 641 – 660.
24. Larone D.H., Medically important fungi: a guide to identification. Washington DC: ASM Press (1995).
25. Larone, D.H. Medically Important Fungi: A Guide to Identification, 2nd ed. American Society for Microbiology, Washington D.C., 1993.
26. Longmore, Murray (2004). Oxford Handbook of Clinical Medicine. Oxford Oxfordshinre: Oxford University Press. ISBN 0-19-852558-3.
27. McInnis M.R., Laboratory handbook of medical mycology, New York: Academic Press (1980)
28. Micheli, P.A. 1729. Novae platarum genera juxta tournefortii methodum disposita. Floence.
29. Mueller GM, Schmit JP. (2006). "Fungal biodiversity: what do we know? What can we predict?". *Biodiversity and Conservation* 16: 1–5. doi:10.1007/s10531-006-9117-7.

30. Murray P.R., Baron E., Pfaller M., Tenover F., Yolken, Eds., Manual of clinical microbiology, 7th ed. Washington: ASM (1999).
31. Nielsen K, Heitman J. (2007). "Sex and virulence of human pathogenic fungi". *Advances in Genetics*. Advances in Genetics 57: 143–173. doi:10.1016/S0065-2660(06)57004-X. ISBN 9780120176571. PMID 17352904.
32. Pasteur, L. 1861. Memorie surles corpuscles organizes qui existent dans'l atmosphere. Examen dela doctrine des generation spontances. Ann. Sci. Nat. 4th ser.16: 5 – 98.
33. Pugalmaran. M. 1997. Studies on the airborne mycoflora of some indoor environments in Madras city (Tamil Nadu, India). Ph.D. thesis. Univeristy of Madras.
34. Rapini, Ronald P, Bolognia, Jean L, Jorizzo, Joseph L. (2007). Dermatology: 2-Volume set. St. Luis: Mosby. Pp.1135. ISBN1-4160-2999-0.
35. S. Frankel , S. Reitman , and A. C. Sonnenwirth, "Gradwohl's clinical laboratory methods and diagnosis", 7th Ed., The C. V. Mosby Company 2 , 1804 (1970).
36. Sabouraud, R. 1892. Ann. Dermatol. Syphilol. 3:1061.
37. Simon-Nobbe B, Denk U, Pöll V, Rid R, Breitenbach M. (2008). "The spectrum of fungal allergy". *International Archives of Allergy and Immunology* 145 (1): 58–86. doi:10.1159/000107578. PMID 17709917.
38. United States Pharmacopeial Convention. 2007. The United States pharmacopeia, 31st ed., Amended Chapters 61, 62, 111. The United States Pharmacopeial Convention, Rockville, MD.
39. Verma. K.S., Lalitha Shorey. And Shorey, L.1994. Aerobiota of a Vegetable market at Jabalpur.JJ of Env. Biol. 15:4. 325 – 329.
40. Wu TC, Chan JW, Ng Ck, Tsang Dn, Lee MP. Li PC (2008). "Hong Kong Med J 14 (4): 103-9 PMID 18382016.
41. www.fda.gov/Food/ScienceResearch/LaboratoryMethods/BacteriologicalAnalyticalmanualBAM/default.htm.

1- Rules and Guidance for pharmaceutical manufacturers and distributors, 2002. pecification).
2- Checklist of settling plates (QU-04-0003-FRM) (Sampling locations).

Index

B

D

E

F

Q

R

S

T